The Quest for the Perfect Hive

The Quest for the Perfect Hive

A History of Innovation in Bee Culture

GENE KRITSKY

OXFORD

UNIVERSITY PRESS

2010

OXFORD
UNIVERSITY PRESS

Oxford University Press, Inc., publishes works that further
Oxford University's objective of excellence
in research, scholarship, and education.

Oxford   New York
Auckland   Cape Town   Dar es Salaam   Hong Kong   Karachi
Kuala Lumpur   Madrid   Melbourne   Mexico City   Nairobi
New Delhi   Shanghai   Taipei   Toronto

With offices in
Argentina   Austria   Brazil   Chile   Czech Republic   France   Greece
Guatemala   Hungary   Italy   Japan   Poland   Portugal   Singapore
South Korea   Switzerland   Thailand   Turkey   Ukraine   Vietnam

Copyright © 2010 by Oxford University Press, Inc.

Published by Oxford University Press, Inc.
198 Madison Avenue, New York, NY 10016

www.oup.com

Oxford is a registered trademark of Oxford University Press

Library of Congress Cataloging-in-Publication Data
Kritsky, Gene.
The quest for the perfect hive : a history of innovation
in bee culture / Gene Kritsky.
    p. cm.
Includes bibliographical references and index.
ISBN 978-0-19-538544-1
1. Bee culture—History.   2. Beehives—History.
I. Title.
SF524.K75 2010
638'.1—dc22   2009022959

9 8 7 6 5 4 3 2 1
Printed in the United States of America
on acid-free paper

For Jessee

Contents

Preface

I WAS IN HIGH SCHOOL WHEN I FIRST BECAME interested in bees. A wild colony had established itself in the branches of a tree near where I lived in south Florida, and some honeycomb had fallen to the ground. I carefully collected a few larvae, placed them into test tubes from my chemistry set, and spent the next few days watching them develop—first into pupae, then adult bees—all within the confines of those tiny glass tubes.

My fascination continued through adulthood. In 1979, I became a bee inspector and met many beekeepers on visits to hundreds of apiaries in Indiana. As I honed my inspection skills, I grew captivated by the hive. I loved the notion of bee space, the critical spacing between frames and the hive wall that makes modern hives possible. The experience of putting on the veil, firing up the smoker, and opening a hive was thrilling. The smoke, the smell of the comb and honey, the lack of peripheral vision due to the veil—all these sensations contributed to the Zen of beekeeping. When I was experienced to the point where I could stop using gloves, I felt that I was one with these creatures. The fluid motion of removing frames became much like a tai chi exercise. You might say I was stung with a love of bees.

When Eva Crane's *The Archaeology of Beekeeping* was published in 1983, I devoured it. It merged my lifelong interest in Egyptology with my love of bees. Reading the chapters on bee boles—the "holes in the walls" built in Scotland and England between 1250 and 1900 to hold straw beehives—I was amazed. I needed to study them. I longed to head for the Scottish countryside to search for old garden walls that might house previously unrecognized bee boles.

It wasn't until my 2001–2 sabbatical, however, that an opportunity to seek out these bee boles arose. I was in Cambridge that year, transcribing Charles Darwin's research notes for *The Descent of Man* for the Darwin Correspondence Project at Cambridge University Library. After suffering through Darwin's terrible handwriting for a couple of hours, I needed a break, and I decided to take advantage of one of the world's great university libraries and scour the literature on these bee-related niches. I developed leads and began to e-mail archaeologists, historians, and museums in Scotland for sites that might have bee boles. That year, my wife, Jessee, and I made several excursions and viewed over 150 bee boles. In the process we found eight new sets for the bee bole register of the International Bee Research Association (http://www.ibra.org.uk/beeboles/), and the members of the Darwin Correspondence Project heard more about bees than they had ever expected.

Before Eva Crane's death in 2007, Jessee and I were lucky enough to have had numerous occasions to visit with Dr. Crane and her colleague, Penny Walker. Penny had been working with Dr. Crane on a number of bee bole projects, and had contacted us upon hearing about my work. Over lunch, the four of us would share our various adventures in the course of finding new bee boles and verifying the existence of others. Dr. Crane was indeed a once-in-a-generation mind.

On my return from England, I continued my research at the University of Illinois Library. As it progressed, I found that my project—and interest—had started to focus on the history of the beehive. After seeing a magazine discuss the concept of experimental archaeology, I developed my own plans to keep bees in skep hives with glass supers, just like nineteenth-century beekeepers. With John Griffith, my colleague at the College of Mount St. Joseph, I started a multiyear study of these old-fashioned hives. These experiments demonstrated to me the value of some of the old methods and hive designs, and drove my curiosity deeper into the history of the hive.

With the encouragement of Joe Graham, editor of the *American Bee Journal*, I started to organize my observations for this book by writing a series of papers for *ABJ*. Preparing this series became a family activity. During a holiday in the Cayman Islands, Jessee, her mother P. J. Romans, and I spent several hours deciphering the meanings of select passages in a sixteenth-century beekeeping book.

The *ABJ* series focused my thinking about beekeeping and provided the foundation that expanded to eventually become this book. Many people read the manuscript and made valuable suggestions to help complete the final work. I thank Robert Waltz of Purdue University, Ric Bessin of the University of Kentucky, John Griffith of the College of Mount St. Joseph, May Berenbaum of the University of Illinois, and Richard Jones of the International Bee Research Association for their critical readings. Hannah Nadel of the Agricultural Research Service, Amihai Mazar of the Hebrew University, and Siavosh Tirgari of Tehran University of Medical Science all kindly gave me permission to use their photographs. I thank Kim Flottum for permission to reproduce images from the 1923 edition of *The ABC and XYZ of Bee Culture*. I also thank Bill McKnight, Chairperson of the Publications Committee of the Indiana Academy of Science, for his council and support. Finally, I thank

Peter Prescott and Tisse Takagi for their help guiding this book to print.

I have always enjoyed honey, and Jessee and I share a fondness for tasting different regional honeys, but our enjoyment is no longer limited solely to their flavor or sweetness. Indeed, it has been enhanced by the quest for the perfect hive.

Introduction

O N A CRISP SPRING DAY, THE BEEKEEPER OF THE estate walked to a row of niches in the garden wall to check on his bees. They had spent the wet English winter boarded up within the niches, their outside access limited to an opening just large enough to allow a single bee to pass through. When the beekeeper had last checked his hives the previous fall, the straw skeps—those inverted-basket hives so commonly seen in England—were strong and filled with honey. But as he approached it became clear that there was little activity in the hives.

Alarmed, he quickly opened the doors to the niches and found one or two dead bees lying near the entrance of the hives. He grasped the handle of a skep and lifted it, fully expecting to find it empty—but to his astonishment it was filled with honey! But where were the bees? They had simply vanished.

The beekeeper performed the spring maintenance he normally did on hives that did not survive the winter. He replenished the hives with swarms from his surviving hives, and within a couple of years, his bees thrived. Eventually, the strange occurrence of 1782 was forgotten.

❧

Across the ocean and over 200 years later, another beekeeper drove up to her hives to see how they had fared during the winter. The white hive boxes glowed in the morning sun, and the air was still—the first hint that something was wrong. Panicked, she opened hive after hive and discovered that, though they were honey-filled, the bees had all but disappeared. The few corpses left behind gave no clue as to what had happened to the colonies.

This scene, repeated throughout the United States in 2008, shook the very foundations of a fragile American industry. Beekeepers were at a loss. What was causing the bees' disappearance? How could it be stopped? The only response was to give it a name: Colony Collapse Disorder (CCD).

The mass bee disappearance of 1782, though it lacks a name, has symptoms that remind us of CCD. What did the eighteenth-century beekeeper do that helped him quickly recover from his loss? What can we learn from him? In this book I suggest that the solutions to many of apiculture's current problems may be hidden in the forgotten practices of beekeeping's past. The adage "those who forget the past are bound to repeat it" seems apt. As we try to ensure the future of beekeeping—and, along with it, the future of agriculture as we know it—we must remember the history of the hive.

∞

The history of the hive is one of innovation—a quest for the perfect hive. An innovation may mean the improvement of existing technology, the introduction of something new, the modification of a process, or the implementation of an idea. Consider, for example, the evolution of the recording industry. Thomas Edison produced the first recordings on wax cylinders, which wore out quickly and were easily damaged. The delicate nature of the cylinders spurred further innovation, and soon people were enjoying phonograph records. These flat disks were still fragile and liable to scratches, but were generally more durable than the original cylinders. Minor changes were made as records were modified to play at different speeds and for longer periods of time, but they remained essentially the same for decades.

The eight-track tape was introduced as a significant improvement over vinyl records, but cassettes, with their small size and ability to be played in a car stereo, quickly surpassed both records and eight-tracks in popularity. Innovation, however, did not stop with cassettes. They were superseded by the compact disc, or CD, which could hold more information, and, after a few years of cars with combination CD/cassette systems, the CD won out. Now, digital music and the iPod represent the latest wave of new-music technology. In spite of all these changes, some purists passionately advocate the vinyl record as the most preferred listening medium and go to great lengths to find the latest releases on vinyl. That is to say, in the face of all this innovation, the old technology that was popular for much of the past century still persists.

The history of beehives follows a similar arc. Hives did not evolve by the simple introduction of new kinds of beehives. After beekeeping became an occupation, creative beekeepers tinkered with hive design to maximize their honey harvest. Some invented totally novel hives, which in turn changed how bees were managed, leading to yet more innovations. Some past designs, despite yielding large honey harvests, have become obsolete, whereas the simple straw skep, the inverted basket that has been used for over 1,500 years, remains in use. Where did this skep—an innovation made during the first millennium—come from, and why does it persist? Why did other seemingly successful hives drop out of favor?

The evolution of the hive is a product of both art and science. Some hive modifications were incorporated to make the hive more ornate and improve the aesthetic appeal of the garden, while other changes were intended to encourage honey production.

Beekeeping is practiced all over the world; however, this book will focus on the innovations made in the management of the European honey bee, *Apis mellifera*. Other bee species, such as the Asian *Apis cerana*, and the stingless *Melipona* species of tropical America, have also been domesticated, but it is *Apis mellifera* that has successfully been introduced to parts of the world outside of its natural range of Europe and Africa (Crane 1999). The story of this bee and its hives takes us from the ancient Egyptians to medieval European castles, through the Industrial Revolution and on to the present.

Today, honey bees are in trouble. These valuable insects, so critical to $16 billion worth of food production, are suffering from mites, diseases, the large-scale use of pesticides, and Colony Collapse Disorder (CCD). The history of beekeeping may provide clues that could help beekeepers and researchers as they struggle to save honey bee populations. Beekeepers will have to build upon this history of innovation, of successes and failures, of art and science, if they want to save not just an industry, but a way of life.

CHAPTER 1

Honey Bees and the Origin of Beekeeping

HONEY BEES AND HUMANS HAVE BEEN CROSSING paths for millions of years (figure 1.1). Indeed, our early ancestors likely raided wild bee colonies for honey, much as chimpanzees do today (Hicks et al. 2005). Both humans and honey bees originated in Africa, each species migrating out of the continent on several occasions. It was one of these migrations—when honey bees left their tropical ancestral home for the higher northern latitudes—that resulted in the honey bees' evolving large colony sizes and building up excessive stores of honey to survive the colder winters, which opportunistic humans noticed and manipulated.

In addition to living in close proximity to humans, honey bees evolved traits that essentially preadapted them for domestication. First, honey bees inhabit cavities. These range from simple spaces protected by overhanging rocks to hollow tree trunks, but this dependence on a stable base of operation is extremely convenient for keepers, that is, for humans. Once in these spaces, the bees build elaborate combs in which to rear the next generation and produce and store honey, the sweet liquid our evolutionary ancestors robbed in the first steps in the evolution of beekeeping.

FIGURE I.I The honey
bee, *Apis mellifera*,
gathering nectar and
pollen. Photograph by
Jessee Smith.

Next, bees' swarming behavior makes them relatively easy to
capture and transfer to a hive, and humans found they could
move them to a more convenient location if desired. Honey is a
high-calorie food that can be stored for later use, while beeswax
is prized as raw material. Finally, bees are hardy creatures that
evolved to thrive in high latitudes and cool climates, and their
stinging behavior provides some deterrence against theft.

The intentional provision of a cavity in which bees can build
comb, lay eggs, and—most importantly—produce honey is what
we call beekeeping. The structure that houses this cavity is
known as the hive. When honey bees are provided with an ideal
hive, they are able to produce more honey than they need to
survive, leaving a surplus for the beekeepers. Thus beekeeping,
from its very nature, is fundamentally and inextricably linked to
hive design.

WHERE DOES HONEY COME FROM?:
BASIC BEE BIOLOGY

Honey bees are social insects with a caste system of individuals.
In each colony there is one reproductive female called the queen
(figure 1.2a), whose reproductive structures mature as the result
of a special diet fed to her during the larval stages. After she
emerges from her queen cell, which is larger than the cells that

produce the sterile female workers, she leaves the colony and embarks on her only mating flight. She mates with several males during this flight, storing the received sperm in the spermatheca, a structure housed in her abdomen. The spermatheca can nurture and feed sperm for three to five years, ensuring that the queen will not need to mate again for the rest of her lifetime. At the peak of her fertility, a queen bee can lay up to 2,000 eggs per day. Fertilized eggs develop into female worker bees, and those that are not fertilized into males called drones (Shimanuki et al. 2007).

Worker bees (figure 1.2b) care for the queen's progeny, clean the hive, search for food in the form of pollen and nectar, and make honey. Those born in the summer during peak honey flows may live only about six weeks, literally working themselves to an early demise. Fall bees, by contrast, may live for four to nine months. The role workers fill in the colony is determined in part by their age. When a worker is newly emerged, she remains within the hive, cleaning cells, feeding larvae, and producing comb and honey. As she matures, she will become a foraging bee in search of nectar and pollen (Shimanuki et al. 2007).

Workers also serve as a first line of defense, protecting the hive from stinging invaders such as wasps. Only female honey bees can sting; the stinger evolved from an ovipositor that functioned as an egg-laying structure in the bee's distant ancestors. The worker's stinger is barbed (a unique characteristic that is

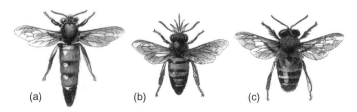

(a) (b) (c)

FIGURE 1.2 The honey bee castes: (a) queen, (b) worker with mouthparts extended, (c) drone (Neighbour 1878).

shared with only a few genera of wasps), and after she stings a victim the stinger and accessory glands are ripped out of her abdomen, resulting in her eventual death. Unlike those of the workers, the queen's stinger is not barbed, enabling her to sting more than once. This is a valuable evolutionary adaptation, as the colony cannot risk losing a queen. A queen bee may sting a potential rival if she discovers her in the hive (Shimanuki et al. 2007).

The male drone bee's (figure 1.2c) only function is to mate with a queen. Generally, a healthy hive has far fewer drones than workers, with drones accounting for only 5% of the bees in the hive. In the fall, the workers expel the males to conserve food.

The products of the worker bees, namely wax and honey, are what have always attracted the attention of beekeepers. Wax is produced by young worker bees before they become forager bees. Four pairs of glands on the underside of the bee's abdomen (figure 1.3) secrete small wax scales. Bees manipulate these scales with their mandibles to form the hexagonal cells of the honeycomb, where the queen deposits her eggs and the workers store honey and pollen (figure 1.4) (Shimanuki et al. 2007).

Honey is the result of nectar collecting by foraging worker bees who leave the colony in search of flowers. The bees are

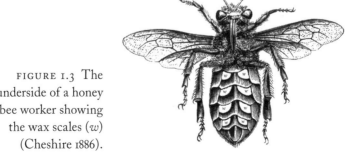

FIGURE 1.3 The underside of a honey bee worker showing the wax scales (w) (Cheshire 1886).

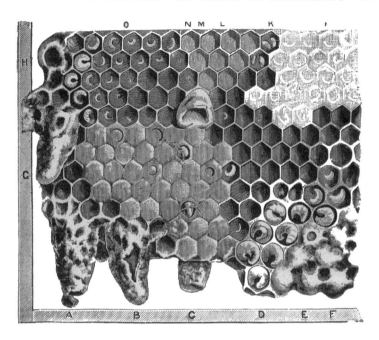

A, Queen Cell, from which Queen has hatched, showing Lid ; B, Queen Cell torn open ; C, Queen Cell cut down ; D, Drone Grub ; E, Drone Cell, partly sealed ; F, Drone Cells, sealed ; G, Worker Cells, sealed ; H, Old Queen Cell ; I, Sealed Honey ; K, Fresh Pollen Masses ; L, Cells nearly filled by Pollen ; M, Aborted Queen Cell on Face of Comb ; N Bee biting its way out of Cell ; O, Eggs and Larvæ in various conditions.

FIGURE 1.4 Honeycomb showing queen cells, brood cells, drone cells, empty cells, and sealed honey (Cheshire 1886).

drawn to the nectar by its smell, and they lap it out of the flower's nectary with their mouthparts. This nectar theft is beneficial for flowers, which evolved nectar to entice the bee (and other pollinators) to visit, aiding pollen transfer from one flower to another and providing plants with increased genetic variation.

Once the foraging honey bee consumes the nectar, she mixes it with enzymes produced in special glands in her head and thorax. These glands start converting the nectar into honey as the bee flies back to the colony. Back in the hive, the foraging bee regurgitates a nectar-and-enzyme droplet and feeds it to a younger worker bee, who adds more enzymes by moving the droplet in and out of her mouth several times before placing it in a cell on the comb. This action removes excess water from the mixture. After comb cells are filled with nectar and enzymes, the workers ventilate the hive to further reduce the water level of the honey through evaporation. When the honey ripens, the cell is capped over with wax and stored until needed (Shimanuki et al. 2007).

Honey is a very stable product. The removal of excess water thickens the honey and increases its sugar concentration, resulting in high osmotic pressure. Osmosis is the movement of water across a membrane to equalize the concentration of substances dissolved in surrounding water and inside any membrane-enclosed structure, such as a bacterial cell. The high osmotic pressure of honey draws water out of bacteria, inhibiting long-term survival. This antibacterial property of honey makes it a good first aid for wounds and burns (Shimanuki et al. 2007).

Though honey makes up a significant part of their diet, bees need more than honey to survive. They also collect pollen from the flowers they visit, shaping it into balls they then carry back to the hive in pollen baskets on their hind legs. The bees pack the pollen into cells on the comb, where it will be kept as a source of proteins and lipids, the class of organic molecules that includes oils, fats, and steroids.

Honey bee colonies are essentially superorganisms. Individual bees reproduce, but colonies split to produce new colonies as well. This divisible quality is another reason that bees are so readily domesticated. The process occurs in the spring, when colonies become crowded and workers start to build queen cells. A few days before the queen and up to 70% of the workers leave in a swarm, the queen lays eggs in the queen cells, and some workers engorge themselves with honey that will be used to feed the newly split colony. After the swarm leaves the colony, the bees usually gather in masses, some as large as basketballs, on tree branches. Here the main cluster will remain for a day or two until foragers return with information about new nesting locations. Meanwhile, back in the original colony, the remaining workers feed the developing queens (Shimanuki et al. 2007). Beekeepers take advantage of this swarming behavior as an opportunity to increase their number of hives by providing empty hives nearby for the swarming bees to occupy.

Beekeeping would not be possible if honey bees had not evolved as a cavity-inhabiting insect that builds up stores of honey as an insurance for its future survival. The structure of the comb, its placement in the cavity, and the bees' swarming behavior provided observant humans the clues as to how to manipulate the bees for their liquid gold.

EARLY BEEKEEPING:
THE HORIZONTAL HIVE

The identity of the first beekeeper is forever lost in time. The earliest evidence of human interaction with bees may date back 8,000 years, to a Mesolithic cliff painting that depicts a human figure robbing a colony of its honey (Crane 1999) (figure 1.5). Honeycomb theft was likely the reason for our ancestors' first intentional encounter with bees, and honey hunting—the purposeful searching for colonies of bees to rob—a logical next step. Though we have no way of knowing the intermediate steps between hive robbing and true beekeeping, historical clues lead us to believe that the ancient Egyptians were the first to engage in beekeeping as we know it.

The first evidence of beekeeping dates from the Fifth Dynasty of the Old Kingdom—the time of the pyramid builders. South of the Sphinx and the pyramids of Giza in the Sun Temple of the Pharaoh Ne-user-re, German Egyptologists found a small bas-relief that depicts Egyptian beekeeping (Kritsky 2007). The relief,

FIGURE 1.5 The oldest record of humans interacting with bees dates back to the Mesolithic period (7,000–8,000 BP). The rock painting, found at Bicorp near Valencia, Spain, shows a honey hunter robbing a nest on the side of a cliff (Kritsky and Cherry 2000).

now in the Egyptian Museum in Berlin, contains four small scenes that illustrate the removal of comb from hives and the extraction and preservation of honey (figure 1.6).

Bees were considered extremely valuable by the Egyptians, and figure quite prominently in their mythology. According to Egyptian myth, when the sun god Re wept, his tears turned into bees. Other deities—Min, Amun, and Neith—were associated

FIGURE 1.6 The oldest record of true beekeeping is from the Sun Temple of Ne-user-re. This bas-relief shows the harvesting and processing of honey. Reconstruction by the author.

FIGURE 1.7 The hieroglyph of a honey bee is often seen as part of the Egyptian royal titulary. Photograph by the author.

with bees as well. In fact, the Temple of Neith was literally called the "house of the bee." Royal names were often followed by the phrase, "He of the sedge and the bee," referring to the Upper and Lower Kingdoms that constituted Egypt, respectively. The bee represented the delta region (figure 1.7) (Wilkinson 1992).

The Egyptians used honey as food, medicine, and even as an offering; several tombs have been found buried with gifts of honeycomb or sealed jars of honey. Honey was so important to the Egyptian economy, in fact, that from the Middle Kingdom onward, beekeeping was under state control (Crane 1999).

Much of what we know about Egyptian beekeeping comes from a painting in the tomb of Rehkmere, who was vizier during the last nineteen years of the reign of the Pharaoh Thutmosis III and at the start of the reign of Pharaoh Amenhotep II. The painting illustrates in detail what beekeeping looked like at the time of the eighteenth dynasty. In this scene (figure 1.8), two beekeepers work three hives. One man holds a censer, a pottery bowl for burning incense, to produce smoke to quiet the bees,

FIGURE 1.8 A reconstruction of the beekeeping scene from the tomb of Rehkmere based on firsthand observation by the author.

while another man kneels to remove the rounded honeycomb and place it into bowls. A large bowl holding a mound of material the same color as the combs sits to the left of the hives and between two other men working with tall, larger vessels. Above these vessels are other containers, one clearly a pitcher for pouring. To the left of this scene, two kneeling men seal containers that are commonly associated with offerings of honey.

The scene is the first to depict the practice of using a censer to smoke the bees. Moreover, the actions are essentially the same as those shown in the Ne-user-re relief, with additional detail. The men in the tomb painting quiet the bees and remove comb, which is shown broken into pieces and mounded up in large containers. This was the initial step in the separation of the honey from the comb. When sections of honeycomb are broken up and placed into a container, the pieces of waxy comb float to the top, leaving the honey in the bottom. The honey is then moved to larger vessels, from which small amounts can be poured into jars.

These early hives looked very different from the hives that we are accustomed to seeing today. They were horizontal, made from mud that was spread out over a mat, rolled into a large hollow cylinder, and allowed to dry. The ends of the cylinder were sealed with disks of dried mud. Bees entered the hive through a hole in one of the disks. This kind of hive can still be found in parts of Upper Egypt (figure 1.9).

Dried-mud beehives placed horizontally on the ground make sense from a design point of view when you consider the natural

FIGURE 1.9 Horizontal Egyptian hives can still be found in parts of Upper Egypt. Photograph by the author.

history of the honey bee in Egypt. Honey bees naturally nest in spaces in rocks in the delta region of the Nile, and a clay or mud tube is a good imitation of these nesting spaces.

Egyptian-style hives spread throughout the fertile crescent. In 2008, Dr. Amihai Mazar and Nava Panitz-Cohen (2008) of the Institute of Archaeology at the Hebrew University of Jerusalem discovered beehives nearly 3,000 years old in the ruins of the ancient city of Tel Rehov, Israel (figure 1.10).

These cylindrical hives, made of straw and unbaked clay, were stacked three hives tall. A straw and clay lid permitted the beekeepers to access the honeycomb for extraction. The area in which the hives were found would have accommodated an apiary of about 100 horizontal hives. As with the ancient Egyptian hives, hives similar to those from Tel Rehov are still found in the region to this day (figure 1.11).

The ancient Greeks produced horizontal hives from fired clay. Closed at one end, the hives were often arranged in the same

FIGURE 1.10 The ancient beehives from Tel Rehov, Israel. Courtesy of the Tel Rehov Expedition, the Hebrew University of Jerusalem. Photograph by Amihai Mazar.

FIGURE 1.11 Traditional hives from Iranian Azerbaijan. Photograph by Dr. Siavosh Tirgari and used with permission.

manner as the ancient Egyptian and Tel Rehov hives: in horizontal stacks of cylinders. The earliest pottery hives date back to 500 BCE. They were grooved on their interiors and equipped with lids that fit tightly within the opening. Each lid had two openings: one small hole in the center for ventilation, and a half-moon opening approximately ¾ inch wide placed at the bottom rim to serve as the entrance to the hive. Other regions produced horn-shaped pottery hives that gradually flared and opened at both ends (Crane 1999). Regardless of the style of hive, they could be placed on the ground singularly or stacked like logs to form walls.

The ancient Romans also saw the value of the horizontal hive, and of adapting hive design to one's environment. Columella wrote that "beehives must be constructed in accordance with local conditions," suggesting that if cork trees were plentiful, hives should be made by cutting a large flat section of cork and rolling it into a cylindrical hive. If cork was not plentiful, hives could be made from a hollow log (figure 1.12), or of woven-together fennel sticks.

FIGURE 1.12 A horizontal hollow log hive (Butterworth 1892).

Horizontal hives in a variety of shapes were used for over four millennia throughout the world. They could be made from easily obtained local materials, and this facilitated their spread into other areas. A sixteenth-century woodcut shows horizontal box hives in use in Italy, and seventeenth- and eighteenth-century woodcuts document the use of horizontal log hives in Denmark and Spain, respectively (Crane 1999). A photograph of soldiers taken in Venice in 1914 (figure 1.13) shows them standing near a shed containing several square and cylindrical horizontal hives.

Horizontal hives are used to this day by beekeepers living in sub-Saharan Africa. Ethiopian hives (figure 1.14) made of dried

FIGURE 1.13 Horizontal hives in Venice in 1914 (Kritsky collection).

FIGURE 1.14 Ethiopian beehives. Photograph by Hannah Nadel, used with permission.

mud or clay look quite similar to hives depicted in the ancient Egyptian tomb of Pabasa (figure 1.15).

In Kenya, cylindrical hives made from rolled bark are hung from tree branches (figure 1.16) to protect them from predation by the honey badger, and to regulate temperature, as temperature extremes are less severe in the branches than on the ground.

Apis cerana, the smaller honey bee species found in Asia, has historically been kept in horizontal hives. In India, this bee is kept in hives made of mud and straw. In Thailand, the hives are fashioned from coconut logs and shaded with palm leaves. Chinese beekeepers also worked *A. cerana* in horizontal log hives. Beekeeping spread into Japan sometime during the first millennium of the Common Era, and it was well established by the twelfth century CE. Little is known about these early Japanese hives, but illustrations of nineteenth-century beekeeping show that horizontal boxes and barrels were the hive of choice (Crane 1999).

The Maya in the Yucatan used horizontal log hives to keep the tropical American stingless bees, *Melipona beecheii*. The logs were stacked on a rack under a shelter, or hung on the side of a hut for protection (figure 1.17). Horizontal pottery hives were used in the Yucatan by the nineteenth century. The use of

FIGURE 1.15 The 26th Dynasty tomb of Pabasa showing the horizontal hives with tapered ends similar to the Ethiopian hives of figure 1.6. Photograph by the author.

FIGURE 1.16 A Kenyan hive hung in a tree. Photograph by Hannah Nadel, used with permission.

FIGURE 1.17 A horizontal log hive typical of Mayan beekeeping. Photograph by the author.

horizontal log hives in Central America developed independently from the rest of the world, but again natural history provides a clue: *Melipona beecheii* nests in hollow trees, so the use of logs is not surprising.

In all these cases, beekeepers tried to create an artificial cavity with local materials that best approximated the honey bees' natural nesting preferences. The key to the origin of beekeeping was observation and providing the bees with what they need to build their comb, produce brood, store honey, and maintain optimal hive temperatures. Yet in spite of their longevity and wide distribution, horizontal hives are primitive, at best, and do not advance beekeeping beyond the practices of the ancient civilizations. To move beyond the ancient methods, beekeepers required a need to change, such as a different environment, which resulted from the spread of beekeeping to the cooler forests of Europe.

CHAPTER 2

The Forest Beekeepers

As with the horizontal hives of the Egyptians, forest beekeeping started with man-made imitations of naturally occurring beehives. Bees often take residence in the hollow spaces found in trees, and this made such inhabited trees prime targets for honey hunting. Honey hunting, a practice that continues into this century, involves carefully observing foraging bees and following them back to their colony. The honey hunters eventually began to supplement their harvest by placing filled honeycomb in areas where it would be found by bees, creating foraging sites of their own (Crane 1999). This practice marked the transition from simply taking advantage of bees' natural behavior to manipulating it for human use, and paved the way for forest beekeeping.

The honey hunter modified trees to make them both more attractive to bees and easier for beekeepers to work. In the early days of honey hunting, whole bee trees were cut down and the honey and wax extracted. But this method was labor intensive, and finding another bee tree was never guaranteed. Honey hunting evolved toward beekeeping as honey hunters started, instead, to cut small holes into the trees to rob honey, leaving the tree intact for future use (and robbing). Cutting the tree usually involved making a long rectangular door that could be removed to get to the comb. When it was closed up, the tree was marked on the outside with a set of cuts to indicate ownership. The forest beekeeper would check his bee trees regularly during swarming season to see if any had been occupied.

This evolution from bee hunter to forest beekeeper can be traced morphologically by the Russian language. The Russian word for hive, *bort*, also means "hollow tree-trunk," and the word *bortnik* translates, literally, to "keeper of wild bees" (Galton 1971).

Forest beekeepers developed a number of unique tools to facilitate their work (figure 2.1). Ladders were obviously useful, but were inconvenient when traveling long distances through

FIGURE 2.1 Forest beekeepers used ladders and hoists to reach the hive opening, bags for the collection of combs, and the bow and arrow to control the forest beekeepers' greatest pest and threat— bears (Bessler 1921).

FIGURE 2.2 Log hives placed on the ground for ease of use. The hive on the left has been fitted with doors to allow for easy access to the interior (Bessler 1921).

the woods. A plank of wood attached to a rope was more compact, and could be thrown over a branch to hoist the beekeeper up the tree. Other special tools included bags or baskets to carry honeycomb, axes to mark the bee trees, and pipes to provide smoke to quiet the bees (Crane 1999).

As situations changed, forest beekeeping evolved. When hollow trees became scarce, hollowed-out logs were hung in trees to provide bees with additional hive space. Eventually, these log hives were moved out of trees and placed in apiaries, groupings of hives in a single area (figure 2.2).

Despite the later inventions of straw and box hives, forest beekeeping endured. When beekeeping was introduced to the Americas in 1622, the straw skeps used to transport the bees were quickly replaced with hollowed logs called "bee gums," called such because they were generally made from the hollow trunks of gum trees (figure 2.3). As in the Old World, American forest

FIGURE 2.3 A bee gum on display at Purdue University. Photograph by the author.

beekeeping was practiced in areas where there was an abundance of trees. T. B. Minor extolled the value of these log hives in his 1857 edition of *The American Beekeeper's Manual*:

> Every one, I presume has seen hives made from hollow trees, by cutting off the log of a suitable length, and then nailing a board on the opening at the top. This is a much better hive than those made of straw. These log-hives are called "gums," in some parts of the country. I recommend this kind of hive to those who wish to keep bees *without any expense whatever*. There is no principle of the habits and economy of the bee, that conflicts with log-hives;...The log-hive is preferable to many patent hives now in use; and I can name several of them that I would not as soon use as the hollow log if compelled to use either.

In addition to being cheap, forest beekeeping was successful, and led to a number of important innovations in log hives. One elaborate tree hive, known as the "Improved Polish Hive," (figure 2.4), was described in 1842 by John Wighton. Wighton believed that some hives of the early nineteenth century had failed because the swarms were difficult to control, and because cold English winters had killed off many of the colonies. He solved these problems by merging log hives, which were common in Poland, with some of the innovations found in hives of the early nineteenth century. The log hive was successful, according to Wighton, because it more closely approximated honey bees' preferred conditions.

The Improved Polish Hive was constructed from the root end of a fir tree that was nine feet long and three feet in circumference. After the top of the tree and the branches were removed, its center was hollowed out to within three feet of the bottom by cutting a section out of the trunk and trimming away the center

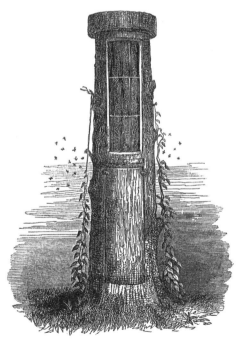

IMPROVED POLISH HIVE,
UPPER DIVISION OPEN.

FIGURE 2.4 Wighton's Improved Polish Hive (Wighton 1842).

wood, keeping the bark and two inches of wood to be reinserted as upper and lower doors to the hive. The inside of the hive was trimmed to form a space seven inches square, which was separated into an upper and lower section by a sliding wooden panel. Four thin strips of wood were nailed to the top of each section to act as bars to support the comb. Two entrances were placed opposite the doors in the lower section, with one at the top and the other at the bottom. The hive was also fitted with glass windows to permit observation of the bees. To keep the tall structure from toppling over, the base was buried two feet into the ground, while the top of the hive was covered with zinc and capped with a circular wooden block.

By reducing the interior of the Improved Polish Hive to seven inches, Wighton believed the bees would start to fill their combs sooner, being warmer because the cluster was more confined. He recommended that bees be kept in the upper division during the winter months, then forced into one of the two sections to make honey harvesting safer. To collect honey, the beekeeper could work the hive on a warm day when most of the bees would be out foraging. All that it required was opening the door of the hive and cutting away the desired comb. The bees would then return to build more comb and fill it with honey.

The log design became so common that in 1882 Charles N. Abbott revised Dr. H. C. J. Dzierzon's *Rational Bee-keeping* and gave the log hive the status of the "original hive." Log hives that were set upright were called Ständer hives, while horizontally placed ones were called Lager hives. Upright hives permit bees to store honey above the brood cluster and are useful in cooler temperate regions; the extra layer of honey provides insulation that protects the cluster during the winter. Bees in temperate areas that start making brood early produce stronger hives than bees that produce brood later in the spring, and the insulation provided by the upright design permits bees in colder climates to start producing brood in the late winter.

FIGURE 2.5 The "Manifold hive," made from two logs and capped with boards (Dzierzon 1882).

Ständer hives had walls that were two or three inches thick, with a door cut at either end. The design of these log hives was also adaptable to more advanced beekeeping practices. If the inside walls of the log were made smooth and parallel, the logs could be furnished with fixed bars or movable frames, important beekeeping advances that will be detailed in chapter 10. Dzierzon also noted that two or more logs could be joined to make a larger "Manifold hive" (figure 2.5) and that the functionality of the hive was improved by adding a dividing board to drive a swarm or capture the queen.

Although Dzierzon had a favorable opinion of log hives, he ultimately preferred the box hive. His reasoning was simple economics. A good log could produce one fine log hive, but if converted to boards, that same log could produce enough material for several box hives. Upright log hives thrived in cooler climates in the nineteenth century, however, and their use spread throughout Europe and Asia. They were used in the Canary Islands, Spain, Hungary, Denmark, and Russia. In Poland, beekeepers with a knack for woodworking carved anthropomorphized upright log hives with the mouth of the animal or human figure acting as the bees' entrance.

For all their popularity at one time, forest beekeeping is now in decline in the Western world. There were considerable governmental efforts during the twentieth century to do away with log hives, particularly in Russia and the United States, because the logs cannot be easily inspected for disease and parasites. This governmental action has taken a major toll on forest beekeeping. But forest beekeepers had already introduced a major design innovation that changed beekeeping forever: the upright hive, which later evolved into the skep.

CHAPTER 3

Skeps

L OG HIVES, FOR ALL THEIR DURABILITY, HAD SOME major drawbacks. They were heavy and required valuable limited resources to produce. A tree of the size necessary to produce a single log hive could provide several wooden boards for building construction. What was needed was a strong but lightweight hive that could be easily produced from inexpensive, readily available materials. The answer emerged from northern Europe, in what is now Germany. The skep (figure 3.1), a hive woven out of plant materials, would become a symbol of beekeeping for centuries.

The first skep hives date from 0–200 CE and were found in Saxony, in eastern Germany. They consisted of woven wicker baskets sealed with a mortar of dung and mud (Crane 1999). The more lightweight straw skep probably originated with the Germanic tribes west of the Elbe River, spreading into Lower Saxony and England by 500 CE (Crane 1999). Unfortunately, because the oldest known straw skep dates back only to 1200 CE and the earlier literature does not differentiate the type of skep hive in use, these dates remain estimates.

The English word "skep" may have originated from the Latin word *sceppa*, but the use of the word to describe a basket of a specific size is from the Old Norse. There are other derivations

FIGURE 3.1 A working skep hive at Hodsock Priory near Blyth, England. Photograph by Jessee Smith.

from Norwegian, Swedish, and Danish, as well as obscure links to Middle Low German and Dutch. Such extrapolations suggest that skeps made their way into England via Central Europe, which is in general agreement with the archeological record.

Although it does not help establish a date of origin, the long history of the word "skep" suggests how it came to be used for beekeeping. According to the second edition of the *Oxford English Dictionary*, the oldest use of the term dates back to 1100, when a skep was a unit of measure; the amount of grain or coal that could be held in a basket of a specific size was called a "skepful." A skep was also the name of a particular kind of basket in the 1300s, and its first known usage with reference to a beehive was in 1494, as recorded in a deed: "The same Kateryne shall have free use to go and come by here hive skeppys being within the Meese and Yards." This etymology suggests that a basket of a particular size was simply adapted by beekeepers for another purpose.

Many early English beekeeping books consider the question of which material is better—wicker or straw. John Levett, in his 1634 treatise, *The Ordering of Bees*, wrote of skeps: "Diverse countries have their several fashions, as well for the matter whereof they are made, as also for the manner and form of their making. But in our country (as you say) the hives made of wicker or of straw are principally in use; but whether of them is better I will not peremptorily determine, because I have seen bees prosper and increase well in both: and I hold it not a material part of the well ordering of bees, to use one or the other, yet do I like best those of straw if they be well made, as the warmer and most agreeable to the nature of Bees..."

Charles Butler also considered the two materials in his book, *The Feminine Monarchie* (1634), but was more definite in his recommendation: "All things considered, the straw hives are better: especially for small swarms."

A century and a half later, in *The Antient Bee-Master's Farewell* (1796), John Keys proclaimed, "Of all such hives as are to stand unsheltered by a house those made of straw are much to be preferred as best defending the Bees both from excessive heat, and excessive cold." The bees' winter survival in straw hives was an important point. Levett (1634) noted that bees and wasps often overwinter in the thatched roofs of houses. Thus, straw must naturally be the superior material.

Straw skeps had many advantages over wicker ones, especially if they were well made. They were lighter than wicker skeps and provided better protection against the elements. Wicker skeps had to have a dung-and-mud mortar applied to the outside to protect the bees, and this covering would break off over time, exposing the colony. Moreover, straw skeps were more durable. Crane (1983) reported seeing skeps that had been in continual use for 150 years, a longevity that later wooden box hives would have difficulty matching.

Preferences varied by country, however, and Levett was not exaggerating when he wrote, "Diverse countries have their several fashions." In Spain, woven hives were constructed from sticks, in Serbia beekeepers used wicker and vines, and in Hungary wicker was used, sometimes thickly coated with mud. The light weight of skeps permitted their transport throughout Europe, and the distribution of traditional hives used in Switzerland documents its neighbors' influences, in addition to revealing geographic preferences. Straw skeps were the predominant hive along Switzerland's northwest border with France, while both wicker and straw skeps were common along the border shared with Germany. Horizontal and upright log hives were preferred along the western Italian border, while, further east, upright bark and log hives were the norm (Crane 1999).

The skill needed to make a tight, well-constructed straw skep was specialized enough that it created a new occupation, the skepper (figure 3.2). The *Oxford English Dictionary* records the term "skepper" as first being used in 1499.

To make a skep, the skepper needed straw and briars or cane for raw materials, a "girth" made of horn or a tube made from leather, and a chicken-bone awl. The girths worked like our modern pasta measurers, providing a template for a consistent thickness to the straw bundle. Once the size was selected, the bundle was twisted to maintain its round shape as a binding of briar was wound about it. The straw bundle, now shaped into a length of round cord, was bound to adjacent cords, with the awl used to pierce a space between the cords through which the briar was fed (figure 3.3). The cords would first be tied together in such as way as to make the bell-shaped top of the skep. Once the diameter of the skep reached the desired size of approximately half a bushel (sometimes measured against a hoop), the skepper

FIGURE 3.2 A skepper and his sons at work (*British Bee Journal* 1893).

FIGURE 3.3 The first whorls of a skep. Photograph by the author.

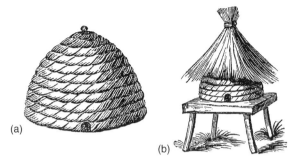

FIGURE 3.4 (a) a traditional skep; (b) a hackle placed over skep (Bagster 1838).

would change the angle of the bindings to make the straight sides. The entrance to the skep was cut by the beekeeper and varied to suit individual tastes. Many skeps had the entrance at the base, while others were nearer the top.

Some of the earliest straw skeps were made in a bell shape with a handle at the top (figure 3.4a). In 1634, Butler wrote, "A handle at the top of each hive is requisite for two uses; carrying the hive, and staying of the hackle." Straw hives had a bowed handle, in contrast to the straight handle found on wicker hives. The "hackle" (figure 3.4b) was a covering of thatch that was tied together at the top and placed over the hive for protection. This thatch was replaced throughout the year to keep it from rotting or becoming a nest for invasive insects.

Straw skeps solved all the problems of log hives. They were easily made, inexpensive, and quite strong when completed. Their strength was the result of twisting the straw cord. Generally, a good skep could support the weight of the beekeeper. They were also lightweight, and several could be carried at the same time—a feat unimaginable with log hives.

Beekeepers needed two skeps for each colony: one in which to keep the colony and at least one empty skep for swarms. Thus, demand arose for good skeps to replace the ones sold during the previous year, as beekeepers sought to increase their number of hives. This demand combined with the skill required to make quality straw skeps helped create an apicultural supply industry. Photographs from the late nineteenth century show skeppers going to market with wagonloads of skeps to sell to beekeepers (figure 3.5).

Straw skep hives are commonly thought of as old-fashioned, improved upon only with the transition to wooden box hives. In fact, quite the opposite was true. Straw skeps provided early beekeepers with an inexpensive template that lent itself to modifications that changed beekeeping practices in ways we still see today.

FIGURE 3.5 A skepper with a wagonload of skeps (Herrod-Hempsall 1930). The skeps are stacked together and laid on their sides to make them more stable during transit.

The first English book written solely on beekeeping dealt with bees kept in skeps. Edmond Southerne's *A Treatise Concerning the Right Use and Ordering of Bees* was published in 1593 (figure 3.6). All we know about Southerne is that he was a gentleman whose book was sponsored by the wife of John Astley, Master and Treasurer of the jewels of Queen Elizabeth I. Southerne was not the first Englishman to write about bees and beekeeping, but his book influenced most of the authors of English beekeeping books that appeared in the seventeenth century. Southerne's

T reatiſe concerning
the right vſe and ordering
of Bees:

Newlie made and ſet forth, according to the
Authors oẁne experience: (ẁhich by any
heretofore hath not been done)

By *Edmund Southerne* Gent.

Better late then neuer.

Imprinted at London by *Thomas Orwin* for *Thomas Woodcocke,*
dwelling in Paules Churchyard at the ſigne of
the blacke Beare. 1593.

FIGURE 3.6 The frontispiece of Southerne (1593).

writing detailed the practices that skep beekeepers employed, which encouraged others to keep bees. For us Southerne's book serves as a time capsule to actions that skep beekeepers used, and how bees were kept in skeps is critical in understanding this period in hive development.

Southerne encouraged the keeping of bees not only for their raw materials but also because beekeeping was a profitable occupation. Beekeeping, as practiced by Southerne, could result in a doubling of the number of hives in just one season. These hives would produce better quality honey and wax because it would be "newly gathered," and could also be traded for wheat, sold outright, or used the following year to increase honey and wax production.

Southerne's specific recommendations, like the use of skep hives, are no longer practiced, but they illustrate problems that still plague beekeepers today. His topics included how to get started, how to maintain the hives, how much honey would be produced, and how to avoid honey bee diseases such as foulbrood. Southerne wrote that the best time to purchase new hives was at Christmas, because they would be cheaper and the straw would be dry and best for use. The best hives would measure fifteen or sixteen rolls of straw high (about half of a bushel) in size. Hives that were too big would take too long to fill with honey and would produce less brood during the second year; thus the bees would take longer to swarm.

As Southerne described, new skeps had to be prepared for use by trimming the insides. This "dressing" of the hive involved trimming away any straw sticking out of the rolls because "when bees are new put into the hive, they would scarce like of it, for that they cannot abide such straws, no more than hairs or any other ragged thing...[T]hey will of themselves bee [sic] so long in their manner in scratching and biting away such paltrie, that they might have filled halfe the hive with Waxe and Honey in that space, if it were done for them." The inside of the skep must also be furnished with crisscrossed sticks to help support the

FIGURE 3.7 (top) The interior of a skep fixed with comb as instructed by Southerne (photograph by the author), (center) one week later, (bottom) one month later (photographs by John Griffith).

weight of the comb (figure 3.7). Finally, a small opening needed to be cut into the hive. The finished hive should be placed on a board or a flat bee-stone (figure 3.8) that was larger in diameter than the bottom of the dressed hive.

FIGURE 3.8 A bee-stone at Pitmedden house (photograph by the author).

A good beekeeper would always have an empty hive ready for a swarm, but sometimes bees would not like these new or previously used hives. Southerne described a procedure that would season these hives for the bees. He recommended that the support sticks be removed and the hive filled with two handfuls of wheat or malt. The malt-filled hive would then be presented to a hog, which would be allowed to eat from the hive while it was slowly turned in the beekeeper's hands "so that the froth which he maketh in eating may remaine in the hive." After the hog had finished eating, the hive's interior would be wiped with a cloth and the sticks put back inside, thus making it ready for the bees. It is likely that enzymes in the pig's saliva would digest some of the complex sugars in the grain, which would then be smeared over the interior of the hive, making it a more hospitable home for the bees.

Southerne felt that the ideal sixteenth-century apiary should be situated with an open southern exposure and, if possible, trees to the north. The hives must be kept three feet apart in order to avoid infection. Foulbrood, a bacterial infection of honey bee larvae, was already a threat to these early hives, and apparently the diseased skeps emitted a characteristic smell. Two foulbrood diseases are now recognized, American foulbrood and European foulbrood. American foulbrood has a distinctive decay odor, whereas European foulbrood has a strong sour scent. It is not known for certain which disease Southerne was describing when he wrote that if "in sommer evenings you walke amongst them, you shall smell a very strong savor issuing out of the hives, which although every hive in himselfe doe not dislike, yet on of another they utterly abhoare, according to the old grosse provervbe: Every mans own filth is sweet."

Southerne's book was an important contribution to beekeeping of the sixteenth and seventeenth centuries. Not only did he provide a wonderful description of beekeeping during Shakespeare's day, he detailed procedures that subsequent writers tested and improved upon, sparking a period of innovation in apiculture in the English-speaking world.

Early skep beekeeping, as practiced by Southerne, required that the bees in a hive be destroyed in order to remove the honey. This was often done by placing the hive over lit brimstone matches, which consisted of pieces of paper that were dipped into molten sulfur and dried. Unfortunately, this resulted in a sullied and smoky-tasting sulfured honey. As beekeepers became more adept at managing their skeps, they realized that killing the bees was inefficient because it reduced the total number of hives. Instead of killing their best colonies, beekeepers could drive their bees from one skep to another.

Driving bees took advantage of the bees' response to smoke, their preference for darkened surroundings, and their tendency to walk upward when the hive was struck rhythmically. Generally, rainy days or evenings were the best times to drive bees, as more of the skep's inhabitants would be inside the hive. The necessary equipment included an empty skep, a smoker, two driving irons, and two skewers. If the skep had a rounded top, a pail or a basket was also required.

The first step in driving bees involved introducing two or three puffs of smoke into the hive, which encouraged the bees to leave the entrance and hive floor and move into the combs to fill their crops with honey. While this was happening, the skep was turned upside down with the apex resting in a pail or basket. Beekeepers took care to rotate the hive in the direction parallel to the comb orientation to lessen the likelihood that the comb would collapse under the weight of the brood and the honey (figure 3.9). Then an empty skep was attached to the original

FIGURE 3.9 A beekeeper driving bees (from a magic lantern slide in the Kritsky collection).

skep by two skewers placed about a hand's width apart. Two larger driving irons were fitted into the two skeps to provide additional support. The entire setup was then rotated to allow light to shine into the full skep, while keeping the empty one as dark as possible. Once the skeps were oriented, the beekeeper would hit the sides of the bottom skep continuously for up to five minutes. The pounding stimulated the bees to run upward. Care was taken so as to not damage the combs of the filled hive. The bees, filled with honey and confused by the overturning of their hive and the constant pounding, rushed into the empty, darkened skep. During this migration, the beekeeper would watch for the queen. If two hives were being merged, one of the queens could be removed. If the newly merged hive was to be requeened, then both would be taken and the new queen introduced (Herrod-Hempsall 1930). This was called the "open" method of driving bees. "Closed" driving involved placing the empty skep directly upon the filled skep and wrapping the sides with a cloth to keep the entire system closed. This method was widely used by less experienced beekeepers or those more afraid of being stung (Cheshire 1888).

By the eighteenth century the bell-shaped straw skep was joined by a flat-topped skep (figure 3.10a). This shape proved

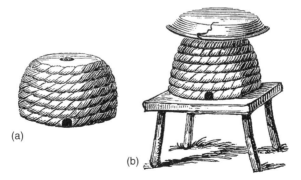

FIGURE 3.10 (a) Flat-topped skep; (b) flat-topped skep with milk pan cover (Bagster 1838).

more efficient because it provided a greater volume of space at the top of the skep for the comb. The handle to which the hackle was tied was eliminated from the flat-topped skeps, which meant that beekeepers needed to find other means of protecting the top of the hive from the elements. An inexpensive alternative was available in the form of cracked (and thus no longer useful) milk pans (figure 3.10b). Keys pointed out in 1780, "For those who can afford it, large earthen milk or pudding pans laid over their straw hives is the best covering that I know of. The largest sort will extend sufficiently to clear the hive-floors of the water that drops from them. Cracked ones may do if the cracks be well spotted with putty, clay, etc. these may be had very cheap...Pans are not liable to [rot and mice] and are more easily removed. Besides which, as our hive floors are not fixed, the pans being heavy, keep the hives so steady that no common winds will displace them."

Skep beekeepers also came up with two methods of increasing the size of the hives, depending on where the additional room was added. "Supering" involved adding new hive space above the established hive, and "nadiring" created new hive space below the hive (Taylor 1880). To accomplish this, Keys (1796) recommended that straw hives be constructed as straw cylinders open at both the top and the bottom (figure 3.11). A wooden grate was placed on the top of the hive to support the new comb. Keys wrote, "They are to have covers of straw bound together in the same manner as the hives; these are to be quite flat, and broad enough to extend half an inch beyond the edge of the hive on which they are to set close and even. They are to be made separate from the hive, being intended to be put on and taken off at pleasure." A cylindrical skep used to create the new space above the hive was called a "super" and the skep that created space below the hive was called a "nadir." Oftentimes an eke, a smaller straw cylinder composed on only four or five bands of straw, was used to increase the space at the bottom of the hive to lessen the likelihood that the bees would swarm.

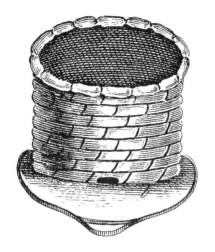

FIGURE 3.11 The cylindrical skep recommended by John Keys (1780).

 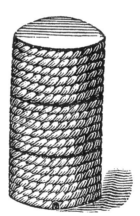

FIGURE 3.12 Storifying using cylindrical skeps and a bell-shaped top skep (Bagster 1838).

Keys was also an advocate of storifying (figure 3.12), or placing full hives upon empty ones. He wrote, "Of all the methods which have hitherto come to my knowledge for the conducting of bees, that of storifying undoubtedly yields much the greatest profit and is the most congenial to their natural habitude, and style of working." Samuel Bagster (1838) wrote that a storied hive will increase thirty pounds in seven days "in a favourable situation and season" compared to a single hive that would increase only five pounds.

Storifying had several advantages. According to Bagster (1838), these included eliminating the need to kill the bees, curtailing swarms during the wrong time of year, reducing the idleness during "lying out," assisting in uniting swarms, providing a defense against mice and bee moths, protecting against bad seasons, and requiring less apiary space. He also wrote that the beekeeper would know when to storify by monitoring the hive's population of bees and its activity.

In addition to increasing the size of skep hives, other beekeepers modified skeps in ways that encouraged high-quality honey production. One such system started with a flat-topped skep with a hole cut in the top, which was plugged with a cork (Bagster 1838). During nectar flows, the hole would be unplugged and a smaller skep would be placed over it (figure 3.13). This would encourage the bees to produce new comb and fill this smaller skep with "virgin" honey. When the smaller skep was filled, it could be removed and sold as is, or the high-quality honey could be removed and the skep reused.

These skep modifications became even more elaborate with the "Preserver" systems (figure 3.14), so called because they saved the bees. Preserver systems included a round skep that was placed upon a larger cylindrical hive called a remunerator. The remunerator had a hole in the top, which permitted the bees, once they had filled up the smaller "preserver" skep, to move down into the larger remunerator to continue comb and honey production. What made the system work was the placement of openings to the hives. When first assembled, the entrance to the hive was through the preserver, which in turn had an opening to the remunerator. In the late fall, when the honey was to be harvested, the preserver was moved in relation to the opening of the remunerator so as to cut off the bees' entrance. At the same

FIGURE 3.13 A method of storifying using a small skep placed upon a board (from a magic lantern slide in the Kritsky collection).

time, an exit was opened at the remunerator's base. This would permit bees to leave the remunerator, but when they returned to the hive, they would enter the preserver opening as usual. After

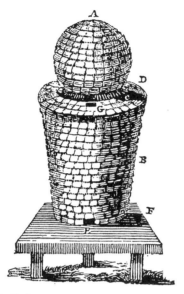

A. The Preferver.
B. The Remunerator.
C. The Floor of the Preferver.
D. The Cover of the Remunerator.
E. The Door-way at its Bottom, to be opened on Depri-
   vation.
F. The Stand.
G. The Aperture in the Cover of the Remunerator at the
   Edge of the Door-way of the Preferver.

FIGURE 3.14 The preserver hive (Herrod-Hempsall 1930).

a day of such activity, the remunerator would be emptied of bees and ready for harvest. This procedure was called "deprivation," and the term "remunerator" was used because the honey contained therein was the payment—or remuneration—to the beekeeper. (Herrod-Hempsall 1930).

The most complex skep system was developed in the late nineteenth century in England and sold under the name "Neighbours' Improved Cottage Hive." George Neighbour and Sons were widely recognized in England as innovators in beekeeping. The "Improved Cottage Hive" was, as Taylor (1880) wrote, "a very much more elaborate structure...than any skep hive hitherto made." The hive (figures 3.15–3.16) combined straw, glass, zinc, cane, and wood into a rather complicated design. The cylindrical straw hive was framed with cane and attached by a hinge to a wooden bottom board that included a landing area for the bees extending from the mouth. The hinge permitted the hive body to be swung open for removal of the comb. Built into the straw sides were three windows, each with a hinged shutter. Behind one of the windows was a thermometer to permit the beekeeper to monitor the temperature of the hive. The top or

FIGURE 3.15 Neighbour's Improved Cottage Hive (Neighbour 1878).

FIGURE 3.16 A Photoshop restoration of a Neighbour's Improved Cottage Hive using parts from the IBRA's collection in the International Beekeeping Museum at Eeklo in Belgium. Photographs by the author.

"crown" board of the hive was also made of wood, into which were placed three tapered one-and-a-quarter-inch holes covered by zinc sliders. These holes were designed to control access to three glass bell jars. These jars were essentially glass supers. (Keeping bees in glass is a tricky process, which we'll examine later.) Resting on top of the hive body and enclosing the jars was a straw skep cover, which included a painted zinc ventilator at the top. The ventilator could be pulled upward to increase ventilation or pushed closed to keep the hive warmer (Neighbour 1878).

The jars of this complicated hive each held six pounds of honey and were often covered with green baize bags serving as insulation. The hive was expensive, at £1 15s. (approximately $150 today), but a cheaper version was available without the thermometer or hinged side windows (Hunter 1875). The advantage of this hive was that the cottage beekeeper could simply slide the opening to the jars closed, cutting away the comb, and place another empty jar over the slider. The keeper would then replace the skep cover and slide open the hole, permitting the bees entry into the empty jar. With careful monitoring of the glass jars, the beekeeper could control swarming.

"Neighbours' Improved Cottage Hive" was a wonder of craftsmanship. Made of fine straw, edged with cane, and tightly fitted to the crown board, they were the focus of much attention at major exhibitions. It was expected that these hives would be used inside a bee house, a free-standing structure to protect the hives against the elements (Neighbour 1878). The expense of building a bee house, coupled with the high cost of the hives, kept these complex hives from becoming mainstream equipment for beekeepers. But the traditional bell-shaped or flat-topped straw skeps continued to be used well into the twentieth century, long after these complex skeps had disappeared.

Skep beekeeping was a major improvement over log hives. It changed how beekeepers managed their hives in two fundamental ways: the use of supers or nadirs, and the control of swarming. These are management practices that we still use today. Books about skep beekeeping spread these techniques throughout Europe, which led to experimentation and further hive changes. No longer were beekeepers walking through forests to observe their hidden bee trees. Beekeeping, driven by the success of the skep, was becoming a science.

CHAPTER 4

Bee Niches

STRAW SKEPS, WHILE DURABLE, EVENTUALLY succumbed to the elements if left unprotected. The hackle or milk pan sufficed for a season or two, but long-term protection required the construction of more permanent structures.

A common method of protecting straw skep hives from the harsh winters in many parts of Europe was to place them within a covered niche in a garden wall or a building (figures 4.1–4.3). Thousands of these niches, previously known as "bee boles" (Crane 1983), were constructed, though they were never recommended by experts. This poses the question: why did they become so common? The practice would have required extensive effort on the part of these early beekeepers, and one wonders how it originated and persisted.

Keeping bee hives in recesses in walls was not an invention of the skep-building period. Columella describes the practice in ancient Rome, but using a construction that is open from behind to permit the beekeeper to work with the hives (Forster and Heffner 1969). The more recent niches built in garden walls, some of which date back nearly eight hundred years, are, by contrast, closed in from behind.

A review of early beekeeping literature reinforces the notion that bee niches in walls are not the best place to keep straw hives.

FIGURE 4.1 Fifteenth-century bee boles near West Barnego, Scotland. Photograph by the author.

FIGURE 4.2 Sixteenth-century bee boles at the Signal Tower Museum in Arbroath, Scotland. Photograph by the author.

FIGURE 4.3 Bee boles constructed in 1840 at the Cliffburn Hotel in Arbroath, Scotland. Photograph by the author.

In 1618, William Lawson wrote, "Some (as that Honorable Lady at Hacknes, whose name doth much grace mine Orchard) use to make seats for them in the stone wall of their Orchard or Garden, which is good, but wood is better." He recommended instead the construction of what we now call a bee shelter, a roofed structure left open on the sides.

John Keys of Wales, one of the most influential beekeeping writers of the late eighteenth century, had bee niches on his property, but never specifically mentioned them in his books. Rather, he recommended that hives be placed near a house so that they may be watched, and so that the hives may be protected by walls, buildings, and hedges (Keys 1780).

In 1819, John Howison of Hillend near Edinburgh, Scotland, described his unfortunate experience with bee niches: "When lately building a garden wall, with good exposure for bees, I ordered a number of niches to be made, into which I put hives. These were, however, so much infested with snails in summer,

FIGURE 4.4 A skep on a low bench (Butterworth 1892).

and mice in winter, that I was under the necessity of removing them to a more open situation." Howison's account is noteworthy because his description is the only first-person account of the building of a bee niche.

In 1838, Samuel Bagster wrote of the walled niche, "The most inefficient [method] is to procure a ledge in a wall, or a low bench, perhaps an old chair or similar article, whereon to place the hive" (figures 4.4 and 4.5). Like earlier writers, Bagster warned against walled niches.

British bee niches were constructed from 1200 until around 1900. Crane (1983) examined the frequency of their construction and found that over 95% were built between the sixteenth and nineteenth centuries, with the eighteenth century marking the high point. Oddly, this surge in niche construction coincides with the publication of several books that specifically recommended *against* it.

The increase in bee niche construction during the latter half of the millennium likely has a simple explanation. The problem lies in terminology. There was no widely used term to describe

FIGURE 4.5 A skep on a shelf (Butterworth 1892).

these structures, which were called bee niches, bee boles, bee garths, and bee holes (Duruz and Crane 1953). During my survey of such structures in England and Scotland, I met only one individual, an elderly woman living in Scotland, who knew these structures as "bee boles." She learned the term while taking courses in agriculture. The lack of a widespread term for these structures meant that warnings against them could have been misinterpreted as recommendations for their use.

The *Encyclopaedia of Gardening* by John Loudon (1835) may have been a source of possible confusion. Loudon's writings were widely consulted during the height of the bee niche-building period, and he wrote several books on gardening in the early 1800s. In the 1835 edition of the *Encyclopaedia* he wrote, "The simplest form of a bee-house consists of a few shelves in a recess of a wall or other building exposed to the south, and with or without shutters, to exclude the sun in summer, and, in part, the frost in winter. Bee-houses may always be rendered agreeable, and often ornamental objects." This confusion of a bee house (a free-standing structure) with bee niches (part of a wall or building) may be the key to the beekeepers' misinterpretation. Bee houses were widely recommended during the 400-year period of bee niche construction, and if simple niches in walls were thought to be synonymous with bee houses, then it is understandable why so many niches were produced.

Taking the term "bee house" to mean "niche" was not limited to the nineteenth century. In 1930, William Herrod-Hempsall, the former editor of the *British Bee Journal*, wrote, "There still remain in various parts of the country bee-houses built hundreds of years ago." He then proceeds to describe "bee garths" and the typical bee niches as examples.

Further confusing the issue was the widespread recommendation that apiaries should be protected with walls. In 1634, Charles Butler wrote that a wall was a good place to protect hives. Over a century later, Keys echoed, "The apiary should be defended from the northern and eastern winds, either by buildings, walls, or by close and high bushes."

In a popular gardening book published in 1856, Shirley Hibberd recommended niches or recesses as garden adornments.

> It would be an excellent plan to attach a stall [hive] of bees to the south wall of a gardener's cottage or lodge, with a glass side towards the interior, so that the operations of the bees might be watched from within. The custom of placing them within an arched recess in the wall of the house was one of old Rome, and is still observed in some countries. We look upon this as a very pretty suggestion for a fancy cottage in any style of architecture.
>
> Unfortunately our suburban cottages seldom have the arched recess, although the tenant might construct it if he thought proper, and it would really add very much to the rural tone, indeed to the classic completeness of a country box, if a

fourteen or sixteen-inch wall, filled in with rubble, were run up sufficiently high for architectural purposes, for the sake of leaving an arched recess for a bee-stall on the east or south side.

Thus these niches, confused with bee houses, were considered a nice addition to a garden.

Whatever their origin, bee niches were destined for a short future. During the height of bee niche construction, box hives—which required less protection than straw skeps—were being introduced. As a result, the last bee niches were built around 1900. But the transition from skeps to box hives did not immediately end the use of bee niches. Beatrix Potter's children's book *The Tale of Jemima Puddle-Duck* (1908) contains an illustration of a bee niche with a box hive standing within its protective walls. Even today, at Pluscarden Abbey near Elgin in northern Scotland, the monks still place skep hives in the thirteenth-century bee niches (figure 4.6) to capture swarms.

CASTLE NICHES

The practice of keeping skep hives in niches was not confined to garden walls, and eventually reached loftier grounds. Although castles more often bring to mind images of knights in armor than bee hives, honey bees played a crucial role in everyday castle life. Honey served as the primary sweetener for food and drink and was the basis for mead, the favorite drink of Queen Elizabeth I (Ransome 1986). The wax produced by the castle's hives was used for candles, and even served as currency. The value of both commodities was such that honey was tithed by the Church and wax could be used to pay fines and tributes (Walker and Crane 2001).

The importance of bee products to English royalty and aristocracy dates back to between 688 and 694 CE, when the laws of

FIGURE 4.6 A bee bole in use in 2001 at Pluscarden Abbey near Elgin, Scotland. Photograph by the author.

King Ine of Wessex proclaimed that rents could be paid in honey. During the reign of King Alfred (885–899), beeswax candles were used as a means of measuring the passage of time. Indeed, during Alfred's day the theft of bees was considered as serious an offense as stealing gold or horses (Walker and Crane 1999).

There are no special accounts describing how bees were managed in castles, but a few castles offer evidence from which we can infer the number and size of their hives. There are twelve castles in Scotland, two in England (figure 4.7), and one in Wales that incorporated bee niches in their walls (Walker 1988). What the castle builders called these niches is unknown. A search of terms used to describe castle architecture prior to 1550 failed to turn up any specific word referring to these niches in castle walls (Gee 1984). Moreover, though a book on the castellated structures of Scotland included bee niches in its castle

FIGURE 4.7 The bee boles at Thornbury Castle. Photograph by the author.

plans (figure 4.8), the authors did not know for what purpose the niches had been included (MacGibbon and Ross 1887).

An impressive set of bee niches can be found at Tolquhon Castle in Grampian, in eastern Scotland. Tolquhon was built by Sir John Forbes shortly after 1500. It underwent a major expansion in 1584, completed five years later, at which time twelve bee niches were placed in the southwest-facing wall of the outer court (figure 4.9) (Tabraham 1999). The size of these bee niches, which measure 25 inches high, 25 inches wide, and 20 inches deep, is comparable to those found in other castles. Cathcart Castle in Glasgow had one bee niche measuring 18 inches high, 18 inches wide, and 18 inches deep. Hatton Castle in Turriff in Grampian has seven niches measuring 28 inches high, 23 inches

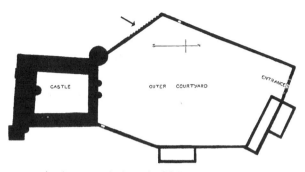

FIGURE 4.8 Architectural plans for Tolquhon Castle showing the bee boles (arrow) (MacGibbon and Ross 1887).

FIGURE 4.9 The bee niches at Tolquhon Castle. Photograph by the author.

wide, and 20 inches deep, whereas Ethie Castle near Arbroath had three niches that were 26 inches wide, 19 inches high, and 19 inches deep (Crane 1983).

Bee niches were added to castles throughout the castle-building period in Scotland. Cathcart Castle was built early in the second millennium, with the bee niches added later. Balbergo

Castle was built in 1569, with bee niches added before 1795, and Ethie Castle's bee niches were in place by 1784.

The niches in all castles were made to be larger than the average straw skep hive, which measured 15 inches in diameter. The excess space in the niche was packed with straw and mud to insulate the hives during the winter months (Crane 1983). Some bee niches not associated with castles included metal hardware for placing doors on the front of the hives to provide some winter protection.

Most castle bee niches were square in shape and placed within perimeter walls, but some styles were built with pleasure in mind rather than the protection of bees. At Midmar Castle in Aberdeenshire, two bee niches were placed on top of each other within a framed and pedimented structure that shares the same artistic finial as the nearby sundial (Fenwick 1976).

The placement and construction of the bee niches was carefully planned. The majority of niches faced south, southeast, or southwest to permit hives to be warmed by the sun (Crane 1983). For most niches, one large lintel stone served as the roof and another as the floor (figure 4.10), and these stones needed to be

FIGURE 4.10 A bee bole at Tolquhon showing the stone layout. Photograph by the author.

carefully selected. If the castle had fortified walls, they needed to be of a thickness that would provide room for a skep hive and yet not be easily broken through if the wall was attacked at the thinnest area.

While bees and their products played an important role in castle society, the vast majority of castles in England, Scotland, and Wales did not have bee niches. Instead castle beekeepers placed their skeps on wooden benches that have long since decayed away. The building of niches in castle and garden walls declined in the late nineteenth century as the improved wooden box hives finally became popular.

CHAPTER 5

Early Box Hives

THE FIRST BOX HIVES WERE SIMPLE HORIZONTAL hives made from wooden boards. Although the date of the first box hive is unknown, a woodcut in A. Gallo's 1596 book on agriculture depicts several box hives—including an upright stack of two boxes and several horizontal hives like those shown in figure 1.13—suggesting that this design existed as early as the late sixteenth century (Crane 1999). These hives did not develop out of a new method of beekeeping, but rather out of the availability of different materials for hive construction.

A novel hive did result from the horizontal box hives. In the late eighteenth century in Slovenia, Anton Janscha equipped a long horizontal hive with holes on the top and bottom boards that could be opened or closed at the beekeeper's will. This permitted the beekeeper to place a horizontal hive on top of another hive, thus expanding the size of the hive and resulting in a stronger colony, which could eventually be split into two hives by simply closing off the openings (figure 5.1). Janscha considered this a form of artificial swarming (Janscha et al. 1900). These box hives were placed like dresser drawers into carts that could be moved from place to place, or they could also be put into bee houses, which will be discussed in chapter 14. These grouped hives could be differentiated from each other by

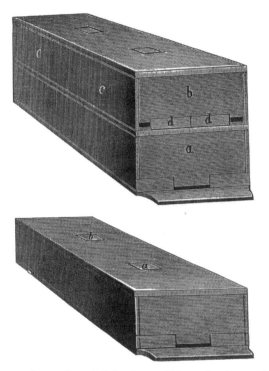

FIGURE 5.1 Anton Janscha's horizontal hives: (top) two hives arranged together to permit the bees to move freely within the boxes, and (bottom) a single Janscha hive with the access holes (a and b) closed off (Janscha et al. 1900).

the front board, which was often painted with a scene from a folktale or religious story (figure 5.2).

A different approach to box hives arose in England, with the transition from straw skep beekeeping to keeping bees inside stacked boxes that began in the seventeenth century. These first tiered hives were octagonal, a design invented by William Mews, who in 1649 constructed an elaborate glass hive that consisted of two octagonal hives stacked one atop the other. The octagon was the shape closest to a circle (like a skep in cross-section) that

FIGURE 5.2 The painted front panel of a Janscha hive. Photograph by the author.

could be constructed from flat pieces of wood or glass. This was thought to be the ideal shape to help bees survive the winter, as they naturally cluster in a circular mass (Thorley 1744).

Christopher Wren, the noted architect of St. Paul's Cathedral in London, modified Mews's hive in 1654, designing a wooden octagonal hive with each panel measuring 15¼" by 9" in size. One side included a window, and another had an entrance that could be controlled by the beekeeper. The boxes fit tightly together, with the bottom of the upper box fitting into the box below it. An opening in the bottom of each box allowed the bees to move from box to box if the beekeeper slid open a pivoting lid (figure 5.3).

The details of Wren's and Mews's hives were published by Samuel Hartlib in his 1655 book, *The Reformed Commonwealth of Bees*. It is likely there that John Gedde, an early proponent of box hives, drew the inspiration for his improved octagonal hive (figure 5.4). Gedde was able to obtain a patent for his hive from King Charles II, and in 1675 published *A New Discovery of an Excellent Method of Bee-Houses and Colonies*.

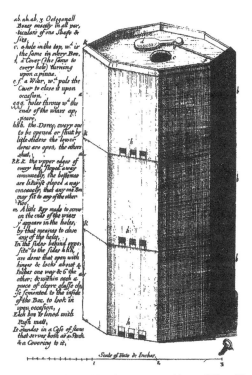

FIGURE 5.3 Christopher Wren's octagonal hive (Hartlib 1655).

Gedde based his octagonal tiered system on five fundamental observations: first, that it was natural for bees to work downward from the top of the hive; second, that bees swarmed because they were too crowded in the hive; third, that the bees' "confusion" about swarming led to their misspending their time "in Luxury"; fourth, that this idleness caused an increase in bees rather than in honey; and finally, that the trouble of dealing with swarms and managing hives discouraged people from keeping bees. In his *New Discovery*, Gedde proposed a new hive that would utilize his observations.

According to Gedde's book, Wren's hive was too large; Gedde recommended that the breadth of each panel should be no more

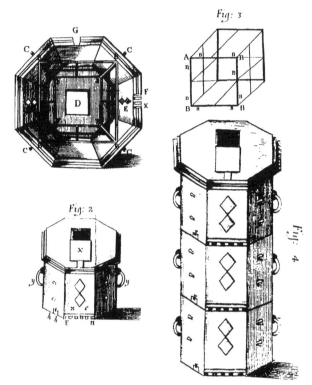

FIGURE 5.4 John Gedde's octagonal hive (Gedde 1675).

than a third of the height and the entire box no larger than a bushel. (Gedde's hive had sides measuring 9″ by 3″.) His book also introduced an important innovation—an internal cubical frame that was inserted into the octagonal box and provided scaffolding for the comb. Though Gedde intended that the entire frame, combs and all, could be removed from the box to harvest the honey, this first attempt at a movable frame was not as successful as Gedde described (Crane 1999).

Gedde was succeeded by Moses Rusden, who built on Gedde's design, albeit with a slightly larger hive. In 1685, Rusden published his own book on how best to use these hives, giving very detailed instructions for the transition from straw skeps to box hives. Rusden told his readers to start by placing a straw skep upon a single octagonal box. The entrance to the skep should be blocked so that bees would be forced to leave via the box's entrance, encouraging the bees to start working the empty box. As the box became full, a second box would be placed underneath, in a nadir system. When the second box was nearly filled, a third box would be placed below the second. At that point the beekeeper could remove the skep and harvest the honey it contained, never needing to use a straw skep again.

To encourage a captured swarm to accept a box, Rusden suggested removing the cubical frame and rubbing the inside of the box with herbs, such as fennel, bean, or elm leaves. Adopting a suggestion from Southerne, Rusden also noted that the best way to prepare a box was to throw some peas into it and let a hog lick them up. After the interior was seasoned, the framework could be reinserted.

This transition from skeps to octagonal hives was accompanied by a change in terminology. Up to this time, a "stock" of bees was a self-contained skep containing a queen, drones, and workers. Rusden (1685) redefined the "stock" as a "company of bees," and made a distinction between a "straw hive" and a "colony," the latter referring to a box of bees. He defined a whole colony as consisting of three boxes, the uppermost used for honey production and the lower two for brood (figure 5.5). Today, we no longer define a colony as including the hive boxes, but rather as a "community of bees," a social unit including the queen, workers, and drones (Shimanuki et al. 2007).

Octagonal hives continued to be used throughout the eighteenth century (Thorley 1744), but underwent a major redesigning with Robert Kerr's invention of the Stewarton hive (figure 5.6) in 1819. Kerr's innovation was the inclusion of nine 1⅛" bars fixed along the top of the box, with sliders (figure 5.7) that could close off the 7/32" spaces between them. The hive

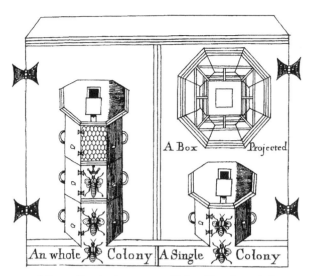

FIGURE 5.5 Moses Rusden's octagonal hive (Rusden 1685).

consisted of three lower "body boxes" and matching upper "supers," which were furnished with seven bars that were wider than those in the body boxes. The boxes were approximately 14" wide internally, with the body boxes being 5½" deep and the supers only 4" deep. Each of the boxes had a window that could be opened to observe the bees' activity, a controllable entrance, handles along the sides, and knobs to tie the boxes together to keep them from being blown over when they were stacked eight or ten boxes high (figure 5.8) (Herrod-Hempsall 1930).

The Stewarton hive enabled the beekeeper to join swarms gradually by placing one swarm into the second box from the bottom and another into the top body box. The sliders were kept closed at first so that the bees had time to get used to each other's scent before contact. After a couple of days, the beekeepers would open the sliders, joining the two swarms into a stronger colony. Because bee colonies do not normally accept two queens, the defeated queen would be found dead outside the hive entrance shortly after the opening of the sliders permitted

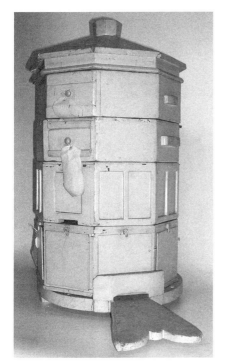

FIGURE 5.6 The Stewarton hive in the IBRA's collection housed in the International Beekeeping Museum at Eeklo in Belgium. Photograph by the author.

FIGURE 5.7 Top view of a Stewarton hive body with sliders in the open position. In the IBRA's collection housed in the International Beekeeping Museum at Eeklo in Belgium. Photograph by the author.

FIGURE 5.8 A tall Stewarton
system (Herrod-Hempsall 1930).

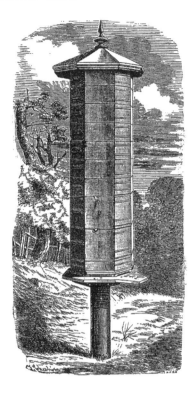

contact. This merged colony would then start building comb
simultaneously in the bottom box and super (Neighbour 1878).

The super had wider bars than the body boxes to promote the
construction of deeper comb for honey. If the queen ventured
into the super, she would find that she could not reach the bottom
of these deeper cells with her abdomen and would not lay eggs
in them; thus, these supers produced considerable amounts of
virgin honey, honey produced in new comb that has never
contained brood or pollen. In 1868, a notoriously bad honey year,
one Stewarton hive filled 10 supers and yielded 200 pounds of
honey (Herrod-Hempsall 1930).

Early editions of the *British Bee Journal* contain several articles on the virtues and problems of Stewarton hive beekeeping.

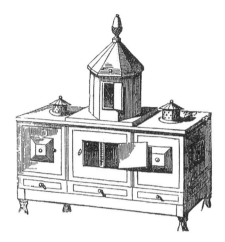

FIGURE 5.9 The Nutt Collateral Hive with ventilators (Neighbour 1878).

They were expensive to build, but those who knew how to use them continued to do so well into the twentieth century.

During the heyday of hive innovation in the mid-nineteenth century, a different box hive enjoyed a brief period of popularity. Nutt's Collateral Hive (figure 5.9) was the brainchild of Thomas Nutt of Wisbech, England, who claimed never to have read any books on beekeeping prior to its invention and felt, therefore, that his hive was a completely original innovation (IBRA 1979).

In fact, collateral hives, which have boxes lined up next to the hive body rather than stacked upon it, were not invented by Nutt. The first collateral hives, from Italy, date back two hundred years before Nutt's design (Crane 1999). Rev. Stephen White constructed collateral hives in England in the mid-eighteenth century and published a book in 1756 titled *Collateral Bee-boxes* (IBRA 1979). Except for decorative elements, Nutt's hives were quite similar to those invented by Rev. White.

Nutt based his hive design on several assumptions about bee biology that were later challenged by master beekeepers. He believed that it was a hardship for bees to walk up a hive carrying heavy loads of nectar and pollen, and that the horizontal orientation of collateral boxes would relieve bees of this burden. Nutt's design also included a thermometer inside the hive with ventilators to permit the beekeeper to cool parts of the hive at will, as he thought that swarming could be managed by careful temperature control in the hive. Finally, he theorized that the purest honey would be deposited in the cooler parts of the hive, whereas the warmer boxes would be filled mostly with brood (Nutt 1832).

Today, there are three collateral hives in the International Bee Research Association (IBRA) collection at the International Beekeeping Museum in Eeklo, Belgium, one of which bears a stamp indicating that it was built using Thomas Nutt's patented design (figure 5.10). These hives consisted of three boxes made of wood, preferably red cedar, that were placed on a single stand (figure 5.11). The central box, which Nutt called the "Pavilion of Nature," was the main hive body, measuring 9" tall, 13½" wide, and 17" deep. It had a tin panel just below the top board that slid out to allow bees to move out of the box through three large holes and into a bell jar. The jar was furnished with an 11"-high octagonal cover with a hinged, pagoda-style lid. The sides of the central box were fitted with tin sliders as well, to control the movement of bees from the "Pavilion of Nature" into the collateral boxes through four openings on each side (figure 5.12). These collateral boxes had the same dimensions as the central box. Nutt designed the hive to have bell jars placed on top of the

FIGURE 5.10 The identification plate on a Nutt Collateral Hive sold by George Neighbour and Sons. From the IBRA's collection housed in the International Beekeeping Museum at Eeklo in Belgium. Photograph by the author.

FIGURE 5.11 The assembled Nutt Collateral Hive. This model was designed to house bell jars above the collateral boxes. From the IBRA's collection housed in the International Beekeeping Museum at Eeklo in Belgium. Photograph by the author.

collateral boxes, but other Nutt hives have been found with ventilators above the collateral boxes instead.

Each of the boxes, as well as the octagonal jar cover, had a door that opened to uncover a window, permitting the beekeeper to check on the activities of the hive. The central box's window opened to a thermometer, which monitored the temperature of the hive. Nutt's hive design was the first to incorporate an interior thermometer (IBRA 1979). Opposite these windows was the bees' entrance to the hive. The hive assembly rested on a base, or hive stand, that was 40½" long, 17" deep, and 4" high; the stand (figure 5.13) was fitted with three drawers lined with tin trays that could be filled with sugar water or honey for feeding the bees in the spring.

Nutt published a book in 1832 titled *Humanity to Honeybees*, which helped spread the word of the success of his hives. In it, he explained how to use the hives, explained the value of taking

FIGURE 5.12 The openings between the "Pavilion of Nature" (right) with the collateral box (left). Note how the tin slider could be used to isolate the collateral box. From the IBRA's collection in the International Beekeeping Museum at Eeklo in Belgium. Photograph by the author.

the hive's temperature, and instructed beekeepers on temperature control through ventilation. To get started with a Nutt hive, a swarm would be placed into the "Pavilion of Nature." Careful observation through the windows would let the beekeeper know when to open the sliders, allowing the bees access to the collateral boxes. To encourage bees to enter these boxes rather than swarm from the hive, Nutt recommended that the inside of the collateral boxes be smeared with honey. This would get the bees accustomed to going into the boxes, and they would shortly begin to fill them with comb. Once the combs were constructed,

FIGURE 5.13 The base of Nutt's Collateral Hive showing the open feeding drawers. From the IBRA's collection in the International Beekeeping Museum at Eeklo in Belgium. Photograph by the author.

a bell jar would be placed on top of the central box, and careful ventilation would keep the collateral boxes cooler than the central box and the jar. After the jar was filled with comb and honey, it would be removed and replaced with an empty jar. The collateral boxes could also be closed off from the central box, and the comb and honey harvested (Nutt 1832).

One danger of Nutt's design was the possibility of removing a collateral box containing the queen. He advised the beekeeper to take care while removing the collateral box and observe the bees' behavior for fifteen minutes after the box had been isolated. If the bees were anxious to leave the collateral box after this period of confinement, then the queen was probably not in the box. On the other hand, if the workers did not leave, then it was likely the queen was inside. In the latter case, the collateral box should be carefully replaced on the stand and the slider to the central box opened. Nutt maintained that the queen would move back to the central box following this period of confinement (Nutt 1832).

Nutt included in his book the details of his 1826 operation as proof of his superior hive design and the methods of its use. In May of that year, he harvested a 12-pound jar of honey and a collateral box weighing 42 pounds, and by the end of the year his total wax and honey harvest was an unbelievable 296 pounds!

Because of their high cost, Nutt's collateral hives never threatened to replace the widespread straw skep hives. George Neighbour and Sons sold Nutt hives for a whopping £5 5s. (Neighbour 1878), about $430 in today's economy. Moreover, the fundamental assumption upon which these hives were based—that it was difficult for bees to walk vertically into hive boxes—was controversial (Taylor 1860). Nutt hives were also colder in the winter than skep hives, resulting in a high winter death rate.

The changing views about Nutt hives were documented in the various books that followed their introduction. In 1860, Henry Taylor, who disagreed with Nutt's assumption that it was a hardship for bees to carry their nectar up into supers, wrote that Nutt's "hives have had their day, where cost and space were not objects." The *Times'* "Bee Master," John Cumming, agreed with Taylor, and wrote in 1864 after his description of Nutt's hive, "I still retain my conviction that the collateral system is not productive. The objection to the storifying system, that the bees have more fatigue in climbing than in traveling on the same level, is not tenable.... The side boxes, also, are too cold." By the 1880 edition of Henry Taylor's *The Bee-Keeper's Manual*, Alfred Watts added, "[Nutt's hive] has wholly ceased to be recommended, and indeed the entire collateral system had all but become a thing of the past." In the end, it was Taylor (1880) who best summarized Nutt's contribution to beekeeping when he wrote, "Although few of Nutt's propositions have been found to stand the test of practice, it ought not to be said that his crude speculations and rash assertions have been altogether without useful results, as they undoubtedly led to farther investigation, and several modern improvements had thus their origin." In other words, innovation was critical to the future of beekeeping.

CHAPTER 6

Beekeeping Comes to America

HONEY BEES WERE INTRODUCED INTO THE NEW World in the early seventeenth century. The first documented evidence of their importation comes from a letter from the Council of the Virginia Company dated December 5, 1621, which listed beehives among the items sent on one of four ships across the Atlantic (Kritsky 1991). The beehives arrived early in 1622, and thus began beekeeping in North America (Crane 1999).

These first hives were most likely straw skeps, as they were the most commonly used hives in England, and beekeeping books describing octagonal wooden box hives had not yet been published. The documentation we have regarding the transportation of beehives on ships was written in the early 1800s, and these procedures also involved skeps (figure 6.1) (Cotton 1842).

Little is known about the kind of hives used in the colonies, but the available evidence suggests that hollow logs (or "bee gums"), barrels (figure 6.2), and boxes were all used up to 1850 (Jones and Wilson 1981). Skeps, inexpensive and popular in England and Europe, did not fare as well in the colonies where straw was in demand for other uses than making hives. J. A. Doddridge (1813), in one of the first American pamphlets

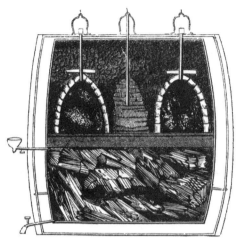

FIGURE 6.1 A hogshead (a barrel or cask with a capacity ranging from 63 to 140 gallons) lined with felt and packed with ice to transport skeps from England to New Zealand. As the ice melted, the water was drained away via the spigots. Such a system would not have been required for the first importation of bees into America, since those bees were transported during the winter and the ship did not cross the equator (Cotton 1842).

FIGURE 6.2 A barrel hive (Butterworth 1892).

on beekeeping, wrote about hive materials, "Straw I should think preferable to wood, but I have never used it for want of a workman to make them of that material." In 1804, T. Minor wrote, "Hives made of hollow trees will do." Of skeps, he wrote, "Straw hives I have never used, and therefore shall say nothing about them." Forty-five years later, T. B. Minor (1849) wrote that straw hives were not much used, and recommended hollow logs for those who wanted to keep bees without any expense.

Box hives were a sign of wealth in Europe, and this status, combined with the continent's ample supply of wood, encouraged the construction of box hives in the States. Minor (1804) recommended pine, but wrote that oak and chestnut would also suffice. An example of an American box hive is on display in the Entomology Department at Purdue University. This simple hive (figure 6.3) consists of two boxes, both open on the bottom. The bottom box is 16⅜" high, 15¼" deep, and 15½" wide, with a 2" by 6⅛"

FIGURE 6.3 A nineteenth-century American box hive on display at Purdue University. Photograph by the author.

rectangular opening in the top. Over this opening rests another smaller box that is 9⅞" high, 9⅝" deep, and 11" wide. The top box, which served as a super, has a circular entrance on one side.

American beekeepers did not remain content with simple box hives, however, and came up with many complex designs that they recorded in the literature. J. A. Doddridge (1813) describes a hive that was essentially a fixed-bar hive "fifteen inches high and one foot square in the inside." The five bars were arranged front to back and were inset into the top edges of the box in a manner similar to shiplap, "so that the tops of them are on an exact level with the edges of the boards." If the craftsman did the work properly, the bars would not need to be nailed onto the box. These bars would be wide enough to allow ½" to ¾" between the hive wall and each bar. At the bottom of the box, there was an entrance 3" wide and "one quarter of an inch or a little more in height." Doddridge recommended supering with two to four boxes, with a board placed over the top super as a cover.

James Thacher, a physician in Boston who accompanied George Washington during the 1776 campaign and an avid beekeeper, described several types of hives in use by 1829. The sheer variety of styles indicates that "Yankee ingenuity" had been hard at work in modifying the box hive. Thacher used a hive built by Dr. T. Willington of West Cambridge. It consisted of a box that was 13½" on each side and 18" high, and inside were two crossbars placed about 6" to 8" from the bottom. Parallel bars, 1" wide and spaced ½" apart, were placed 2" to 4" above the crossbars, creating an interior with two "apartments." In the top "apartment" would be placed nine smaller boxes, each 4⅝" square, and over those boxes would be placed a hinged lid. The bees entered via an entrance in the bottom. Working the smaller boxes first, they would then fill the space below them with comb. The smaller boxes were removed when they were filled with honey and replaced with empty ones. Thacher wrote that this system resulted in the regular swarming of bees, enabling the beekeeper to quickly increase the number of hives, and providing fine comb honey "free from bee bread or young bees, and of the most delicate whiteness, and delicious flavor." The boxes would provide three annual harvests of honey and the comb below would provide ample honey for the wintering of the bees.

Thacher improved on Willington's hive by dividing the upper chamber into drawers or sliding boxes that could be withdrawn from the back of the hive. After the drawer was filled with honey, a steel or brass slider would be inserted beneath it to stop the escape of the bees when the drawer was being exchanged with an empty drawer. Thacher also placed a pane of glass in each of the drawers to enable him to monitor the bees' progress. During 1828, Thacher harvested 38 pounds of honey from one of these hives.

George Morgan originally from Princeton used collateral hives (figure 6.4) modified from the writings of Rev. White. Morgan's hive boxes were made of seasoned wood, and measured 10" square inside. A 7" by 9" window was equipped with a sliding shutter on the back, and two holes were cut on the side of the boxes to permit communication between them. The top hole was 3" long and ½" high, and the bottom hole was 6" long and ¾" high. The entrance, which was 3" to 10" wide and ½" high, could be opened or closed depending on the season. Morgan claimed that each box would produce thirty pounds of honey each season

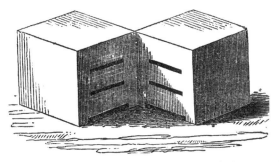

FIGURE 6.4 An early American collateral system (Minor 1849).

and cast out several swarms each year, and that the first swarm would fill another collateral system during that season. When winter approached, the bees gathered in one of the two boxes, leaving the other ready for harvest (Thacher 1829).

The "Charlieshope" hive, attributed to Mary Griffith of New Brunswick, New Jersey, incorporated a number of new features. It measured 13" square at the top, but the sides decreased to 7" wide at the base. The front of the hive was 26" high, but the back was only 20" high. This required that the bottom board be an inclined plane, which was fastened by hinges at the back and held in place by hooks on the sides. The resulting hive was a strange-looking box with a unique profile (figure 6.5). The top was a 1"-thick board that was screwed onto the top of the hive to prevent warping. Three 1" holes were cut into the top, in line with the front of the hive and spaced ¼" apart. These holes were plugged when the hive was not fitted with a super. When the hive was filled with comb, the three plugs would be removed and a super measuring 13" square and 8" to 10" inches high would be placed on top of the hive (Thacher 1829).

On the front and back of the Charlieshope hive were two cleats that held the hive suspended above the ground in an external

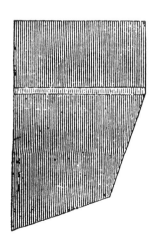

FIGURE 6.5 Profile view of Charlieshope hive with its slanted bottom board. The hive would be held off the ground by a stand as shown in figure 6.6 (Minor 1849).

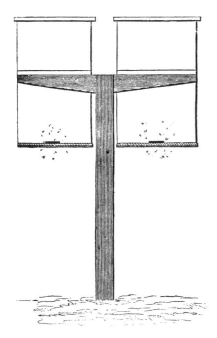

FIGURE 6.6 Frontal view of two Charlieshope hives illustrating the hanging system (Minor 1849).

"moveable frame." (figure 6.6). The frame could be carried from place to place as needed. This suspension system kept the hive away from mice, and the hinged, inclined bottom board permitted easy cleaning and allowed condensation to drip out of the hive.

America was the land of opportunity, and immigrants from Europe who could not afford expensive wooden box hives in their homeland could fashion hives out of wood in the New World. These hives proved an example of their new prosperity in America. Skeps, though never widely used in America, remained in our collective psyche as a symbol of thrift, industriousness, and economy. However, the use of wooden hives by early American beekeepers opened the door for vast experimentation with hive designs.

CHAPTER 7

Glass Jar Beekeeping

To the modern beekeeper, the idea of keeping bees in glass jars (figure 7.1) must seem an odd one. However, between 1750 and 1870, the use of glass jars, especially in small beekeeping operations, was quite common—so common that different beekeepers sold their preferred honey jars and debated the virtues of the various shapes.

Beekeeping in glasses evolved from the practice of supering, providing a means of enlarging the available space for the bees in an effort to control swarming and to increase the honey production of a hive. Its invention is attributed to Rev. John Thorley, whose 1744 book on beekeeping had a great influence on the practice for the next century (Anon 1851). In its simplest form, glass jar beekeeping involved adding a glass bell jar to an existing box hive or skep (figure 7.2) that had been constructed with a flat wooden top with a hole allowing the bees to move from the hive body to the jar. In theory, the glasses could be added as the colony became stronger, and the bees filled them with comb honey (figure 7.3).

Encouraging bees to work jars, however, was apparently not an easy process and required considerable care on the part of the beekeeper. John Milton (not to be confused with the poet), a nineteenth-century proponent of glass beekeeping who sold a variety of hives in London, published a small pamphlet on

FIGURE 7.1 Two glass honey jars, c. 1850, in the IBRA's collection in the International Beekeeping Museum at Eeklo in Belgium. Photograph by the author.

FIGURE 7.2 A straw skep fitted with a wooden top with a hole to permit the use of a glass jar. From the IBRA's collection in the International Beekeeping Museum at Eeklo in Belgium. Photograph by the author.

FIGURE 7.3 A glass bell jar
filled with sealed comb
(Cotton 1842).

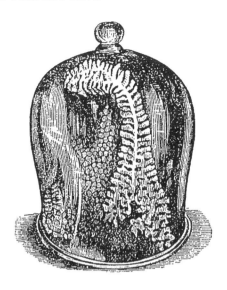

beekeeping in 1823. A copy of Milton's pamphlet is kept in the University of Illinois Library, and glued to its back pages is a letter from John Milton to the booklet's owner, William Wise, who complained to Milton in 1828 that his bees were not working the glass jars in the hives that he apparently purchased from Milton. Milton's response, dated June 28, 1828, reads:

> In answer to your letter which I have this morning received. I am at a loss to know why those persons so very long used to Bees could not satisfy you respecting the probable cause why the Bees have not worked in the Glasses- you say it has *thrown a fine swarm*, the weather previous to the early Part of June or middle of May was not favorable for Honey and of course you could not expect them to obtain Honey in such bad weather = and the next occurrence of *swarming* will *present them*, till the Honey Dews or Lime Trees Blossoms of another *chance* These old Bee Keepers must have made very little enquiring if they think the Bees will not work in Bell Glasses as I cannot Suppose the Bees at Lombridge are different from most other counties. Would they

credit that the Hon[orable] Capt. Cockrane told me last week, that he put a Swarm into a Hive he had of me last Season and that he took 8 glasses from it (that is 4 twice filled before the 4th Sept that the Hive has been very Strong all the Spring and Swarmed last week as his acc[ount] received him by Post last week, This in the first year is *rare* therefore I mention it also and I wish you to observe in the use of these hives you are not debarred from pursuing the old plan of Suffocating—but should nature favor, you have an opportunity of reaping part without taking all = therefore I have only to request you to have patience—I beg to enquire if you have Stopped the glasses round with mortar, are they pretty well air tight, is the light excluded and the board turned to have the holes even if all this is done you have only neglected one part which was to have taken off the glasses, not to let them have remained all the Winter—They collect cold air, Damp, and Smell musty as a consequence but with all these I think your Bees will ultimately work in them—

Your ob[edien]t ser[vant]
John Milton

As Milton's letter attests, using glass jars as a means of supering did not always seem to work.

Glass jar beekeeping required many steps to force the bees to work in the jars, but as proponents such as Keys (1780) wrote, glass jars "are of real use." Keys's *The Practical Bee-Master* included an entire chapter of advice on glass jar beekeeping, detailing the steps required to ensure that the bees would fill the jars with prime honey. Keys felt that jars that held a peck (about two gallons) of honey were best because they would be filled within the limited time of the early honey flow and thus give the best honey, but beekeepers could purchase jars of various sizes and shapes (figure 7.4). In preparing the jar, it was important that it be cleaned, wiped with wax to help the bees walk up the sides,

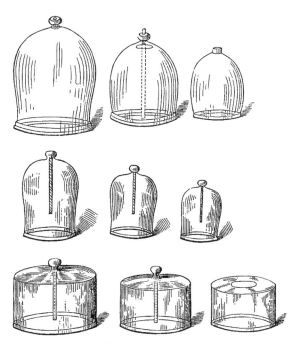

FIGURE 7.4 A selection of popular glasses used in honey production. The top row glasses were sold by Pettitt, the center row sold by Neighbour, and the bottom row sold by Taylor (Cumming 1864).

and furnished with some virgin comb. Many honey jars were also made with a hole in the top into which a small perforated stick could be placed to give the bees a surface in the center of the jar on which to attach the comb (Keys 1780).

Keys recommended first putting a swarm into the jars before using a box or skep. The glass jar would be placed on a board with an opening that permitted the bees to enter and exit, and the whole assembly, covered with a cloth, would be kept in a bee house. If the bees had started to work the comb within two to three days, the beekeeper could be assured that they had accepted the jar. After the bees had filled the jar with comb, the jar would be placed on a skep or box hive to encourage the bees not to

FIGURE 7.5 The author removing a glass super using a wire. Photograph by John Griffith.

swarm but to expand the colony into the extra space. When the jar had filled with sealed honey, it was replaced with an empty jar. Otherwise, the bees would begin to transfer honey to the box or skep hive. If the season was right, the beekeeper could expect two large jars of honey per year from the hive.

To remove the jar, the beekeeper would first slide a knife or wire under the mouth of the jar to separate it from the hive (figure 7.5). The jar would be taken a short distance from the hive, where any remaining bees would be blown or smoked out of the jar and would return to the hive. The jar opening was covered with a cloth and could be stored for up to two years, or could be carried to the table for immediate use.

In addition to recommending jars as a supering mechanism, Keys (1780) included directions for a hive that was constructed almost entirely out of glass jars. This system used seven three-pint glass jars upended on a board that had small holes cut to give bees access to the jars. The board was placed on top of another board, with just enough space left between them to permit the bees to pass from jar to jar, but not to build comb. Cane lined the edges of the boards, with a single opening serving as a hive entrance. The keeper would then place the queen, with one wing cut to stop her from leaving, into an empty jar. Workers were added a spoonful at a time until there were enough bees to work that jar. Workers were then added to the other jars. The result was a glass hive of seven jars interconnected by a common base. After the jars were filled with brood and honey, the glasses and their bottom board would be separated from their base and placed on a box hive. This would encourage the bees to start to fill the box with comb and brood. Glasses were successively removed from the hive after their brood hatched and their cells filled with honey.

As evidenced by the letter from John Milton, getting bees to work glass jars was not an easy process. Along with my colleagues, John Griffith and Megan Cannon, I tested glass supers in southeastern Indiana in 2005. We re-created an 1823 Milton hive (figure 7.6) by modifying a skep to be topped with

FIGURE 7.6 Milton's 1823 cottage hive (Milton 1823).

FIGURE 7.7 A modern reconstruction of Milton's 1823 cottage hive. Photograph by the author.

a flat board (figure 7.7), following instructions found in Neighbour's 1878 book, *The Apiary*. We used bell jars with a ventilation hole drilled into the stem, similar to the glasses I examined in the IBRA collection while it was in storage at the Museum of English Rural Life in Reading, England. Fundamental to Milton and Nutt's philosophy was the notion that keeping the jar cooler than the skep would discourage the queen from laying eggs in the glass jar. To test that idea, we placed HOBO data loggers into the skep and the jar. These data loggers recorded the temperature every five minutes for several weeks.

During the first year, we were unsuccessful at getting the bees to work the glass, possibly because conditions in the Ohio River valley are much hotter and more humid than those in nineteenth-century England. We fared better on our second attempt. By using a smaller swarm and placing the jar, with a piece of wax comb placed inside as a guide, on the skep soon after installation of the swarm, we were able to get the bees to work the jar. Within ten days, the bees had extended the piece of comb and had started to put nectar in the super. Within a month, the bees had added additional comb, building one section upward from the comb through the bung (the hole connecting the jar to the skep). As the comb grew, the bees affixed it to the sides of the jar. As predicted by early beekeepers, the comb in the jar was filled with nectar and virgin honey (figure 7.8), and

FIGURE 7.8 The glass super filled with nectar and sealed virgin honey. Photograph by the author.

no brood or pollen was found within any cell. Our temperature readings also vindicated Nutt's hypothesis that the jars would be cooler than the brood chamber. We recorded over a thousand temperatures during the course of our study and found that the glass jar was significantly cooler than the hive body.

Glass jar beekeeping was popular in Great Britain during the nineteenth century, with awards given at bee shows for the best glass jar filled with honey. They also provided an easy method of removing fine honey from the hive in small containers while letting the beekeeper watch the bees at work. Indeed, larger jars were used as observation hives by those beekeepers who wanted to record the activities of the formerly hidden world inside the hive. But keeping bees in glasses declined as new methods of producing comb honey, virgin honey in clean wax combs, were adopted in the early twentieth century.

CHAPTER 8

The Crystal Palace Exhibition: A Beekeeping Showcase

IN BEEKEEPING HISTORY, THE YEAR 1851 STANDS as a milestone that separates the old and the new. Around this time, people experimented wildly with hive designs and materials. Beekeepers in nineteenth-century America were using an array of wooden boxes or keeping bees in old barrels, primitive bee gums, and log hives. In England, you would find beekeepers using skeps, collateral hives, octagonal hives, and some combinations thereof.

Rev. J. Thorley, an eighteenth-century English beekeeper and author of *Melisselogia or, the female monarchy*, designed a combined system using an octagonal wooden box that had a hinged door on the side, a hive entrance at the base, and a hole cut in the top (figure 8.1). The hole at the top was controlled by a sliding panel made of wood. When a swarm was captured, it was first placed into the wooden box. After the box was filled with brood and honey, a straw skep would be placed over the hole, which was opened by pulling back the slider. Bees would then fill this skep with honey. If the colony was very active and

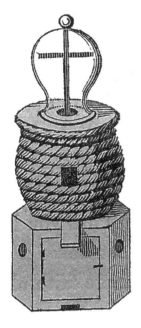

FIGURE 8.1 Thorley's depriving system used a simple flat skep with a bell jar, which was placed on top of an octagonal wooden box (Bagster 1838).

quickly filled the skep with honey, a glass jar could be placed on top of the skep as described in chapter 7.

The combination of skeps and boxes continued into the nineteenth century with the "nether" system (figure 8.2) described by Henry Taylor (1880). The nether was a wooden box that was 6" to 7" inches deep and 11" square. It was made of ½"-thick wood and had a sliding window in the back used to monitor the bees. On top of the nether was a straw skep, which rested on a bottom board with a landing on the front and a small opening in the floor that could be controlled using a tin slider. When the skep filled with brood and honey, the hole in the floor was opened by pulling back on the slider, giving the bees access to the box. Once the box was filled with honey, the slider could be closed, isolating the skep from the nether. The skep could then be lifted off without disturbing the bees in the skep.

With the Industrial Revolution changing the world, the time was ripe for someone to usher beekeeping into a new age. These

FIGURE 8.2 Taylor's nether
system combining a straw
skep with a wooden box
(Taylor 1880).

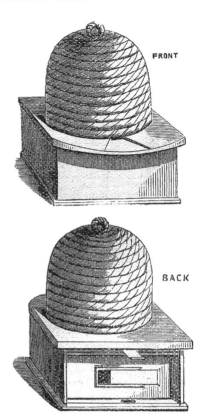

efforts took center stage at the first world's fair, held in 1851 in
London. The event, meant to showcase the political might and
ingenuity of the Industrial Revolution, was held in the Crystal
Palace, an elaborate glass structure built specially for the fair. Of
the 13,000 exhibits, the 60 displays on beekeeping were among
the most popular, and, like the rest of the exhibition, the
beekeeping displays bore witness to a period of innovation and
experimentation.

Two of the most popular apicultural exhibits showcased the
inventory and inventions of England's premier beekeepers, John
Milton and George Neighbour. Milton, an advocate of glass jar

FIGURE 8.3 John Milton's display, showing three of his elaborate hives. From left to right: the Mansion of Industry, the Royal Alfred Hive, and the Milton Unicomb hive (Anonymous 1851a).

beekeeping (Milton 1823), included in his exhibit elaborate observation hives that attracted the attention of the royal family. Milton's Royal Alfred hive (figure 8.3, center), named after Prince Alfred, Queen Victoria's son, was an oddly shaped box with windows that allowed visitors to watch the bees work in the hive. The upper floors of the hive were slanted like a gabled roof, and upon these floors were three large bell jars that, when filled, could each hold 18 pounds of honey. The inclined floor allowed dead bees and other refuse to slide down to the edges of the hive for easy removal (Richardson and Westwood 1852 and Anonymous 1851).

Milton's "Mansion of Industry" or "College of St. Bees" (figure 8.3, left) looked like a mansion or palace with a portico entry and three stories, but this was an elaborate façade that opened to reveal the inner workings of the hive. Inside were four joined swarms that Milton obtained in late July 1850. He estimated that over 200,000 bees worked this hive, and claimed that more than one queen was producing eggs to maintain the large population. This was an especially popular exhibit with children because it was placed low to the ground. Adult visitors had to crouch down to see the hive's inner workings, but the children could observe easily at eye level (Anonymous 1851).

Another of Milton's unusual hives was the unicomb hive (figure 8.3, right), which was constructed like a large picture frame, with just enough space for a single comb and the working bees. It rested on a central pivot, enabling the visitor to rotate the hive to see both sides. At night, the doors to the hive could close off this observational window. One visitor noted that, as it did not have space for a winter cluster, it would only be useful for one season. For those with such practical concerns, Milton showed three other hives that demonstrated more effective forms of beekeeping, including his prize-winning "Improved Cottage Hive," a straw skep with a single glass jar on top (Richardson and Westwood 1852).

George Neighbour and Sons, known for their exhibits at the London Zoo and in Hyde Park (Neighbour 1878), put on a similarly elaborate display (figure 8.4). Neighbour's exhibit included a Nutt Collateral Hive (figure 8.4, center), and his own expensive "Improved Cottage Hive" (figure 8.4, lower right) with its three bell jars under a straw skep cover. His "Ladies' Observatory Hive" (figure 8.4, lower left) consisted of a low, thick, flat-topped glass hive that permitted access to a smaller glass jar placed on top of the flat-topped jar. Covering the entire glass hive was a ventilated skep. Neighbour thought women beekeepers would enjoy watching the activities of the bees inside the hive without the worry of getting stung.

Neighbour's display also included two of his box hives. The single box hive (figure 8.4, second from the right) included a thermometer, shuttered windows on each side, and a gabled top that, when lifted off, revealed a single glass jar. Below the hive was a drawer that extended out to feed the bees. Neighbour's ventilating box hive (figure 8.4, second from the left) featured a

FIGURE 8.4 George Neighbour and Sons' exhibit, showing five hives. From left to right: the Ladies Observatory Hive, the ventilating boxed hive, the Nutt Collateral Hive, the single boxed hive, and the Improved Cottage Hive (Anonymous 1851a).

metal slide that could be pulled away to permit ventilation through a perforated floor (Richardson and Westwood 1852 and Anonymous 1851).

Milton's and Neighbour's exhibits, with their unusual designs and diversity of hives, were designed to draw attention to the science of beekeeping. Other exhibitors, however, demonstrated new hive designs to promote innovative beekeeping. Generally, the hives at the Great Exhibition could be grouped into two classes: hives that could be intentionally increased in size to better control swarming, and hives that could not. In the former class were four methods of increasing size: supering, using a nadir, combining both supers and nadirs, and, finally, using collateral hives to increase size laterally. Hives with fixed sizes were one of three types: hives with movable frames or bars, unicomb observational hives, and traditional straw skep hives (Richardson and Westwood 1852).

FIGURE 8.5 Golding's bar hive
(Filleul 1856).

MR. GOLDING'S HIVE.

Four hives at the exhibition were intentionally designed to enable the beekeeper to remove comb for examination or harvest. R. Golding of Hunton presented two hives using a bar system. His Grecian Hive (figure 8.5) was 9" high, nearly 12" across at the top, and gradually tapered inward toward the bottom. Along the top were placed seven to ten bars, each with a small piece of comb attached to the underside to give the bees a starting point from which to build. The flat-topped cover was made of straw, with an opening in the center for a bell jar (Richardson and Westwood 1852).

Golding's Leaf Hive was constructed of wood with the frames hinged so they could be opened like a book. He also included starter comb within his leaves to indicate to the bees where they should start their work. The design's brass butt hinges permitted the beekeeper to open the hive between any two frames or to remove any one frame at any time. Care had to be taken with this hive, however, because bees would often escape from the adjacent frames when it was opened.

In the midst of all these English designs, one hive exhibited by the Frenchman M. Debeauvois, a physician from Seiches, attracted considerable notice. Indeed, the *Illustrated London News* (Anonymous 1851) urged, "[W]e strongly recommend English bee-keepers to examine these hives for themselves." To a wooden bar hive of simple design (figure 8.6), Debeauvois had simply added a sloping movable roof. Inside the hive were nine four-sided frames, with top bars wider than the sides of each frame to allow for easy removal. Small diagonal strips of wood

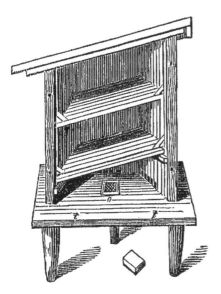

FIGURE 8.6 Debeauvois's movable-frame hive, showing the interior frames (Richardson and Westwood 1852).

kept the frames evenly spaced within the hive. The bottom board included a small ventilator, and there were entrance holes cut along the bottom on each side. The inclined frames and increased space below them permitted the easy removal of dead bees and other refuse. Debeauvois displayed other variations on this hanging-frame system, including a horizontal-framed hive and a straw hive with frames. The ingenuity of these hives won Debeauvois a medal from the Exposition judges for his displays (Richardson and Westwood 1852).

Another English attempt at a movable-frame hive could be found at George Neighbour's exhibit (Anonymous 1851a). Unlike the Debeauvois hive, Munn's bar and frame hive (figure 8.7) had four triangular frames that were hinged on one side and pivoted up into a glass observational chamber. The display advertised that this hive permitted beekeepers to observe the bees and harvest honey without the use of a smoker, but its limited size and more complicated pivot structure seemed highly impractical next to Debeauvois's hive (Richardson and Westwood 1852).

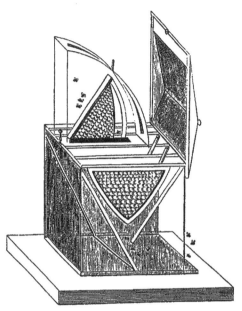

FIGURE 8.7 Munn's bar and frame hive that was displayed in Neighbour's exhibit (Anonymous 1851a).

The 1851 exhibition was a great success and a testament to Western technology. The beekeeping displays illustrated the diversity of practices, from simple skeps to the most elaborate hives of the day. Nevertheless, the interest in the French hive with its movable frames was a sign of things to come. During this period, an American cleric, Lorenzo Langstroth, was quietly working in Philadelphia on a major innovation: a true movable-frame hive. At the time, the exhibitors at the Crystal Palace did not realize that this celebration of apicultural technology was a "last hurrah" for the old cottage and box hives. Beekeeping as they knew it would soon completely change.

CHAPTER 9

Bars, Frames, and the Bee Space

THE MOVABLE-FRAME OR BAR HIVES SHOWCASED at the Crystal Palace exhibition were not the first to feature a system that allowed for the easy removal of an individual section of comb. Golding's bar hive closely resembled a Greek hive that George Wheler described in 1682 (figure 9.1). This hive, like Golding's, was wider at the top than at the bottom, reducing the likelihood that the bees would attach combs to the sides of the hive. Thin wooden strips were laid over the opening to provide building guides for the bees. In theory, the beekeeper needed only to lift the bar out of the hive to remove the comb.

The Greek hive likely evolved between 340 and 260 BCE, when bees were kept in clay pots with openings wider than their bases. The notion is that Greek beekeepers may have placed a board over the pot's opening to keep out the elements. Honey bees able to find an entrance into this protected space would have attached comb to the board, but not to the sloping sides of the pot. When the board was removed, it would have lifted the honeycomb out of the pot as well. Slats of wood separated by small spaces eventually replaced the boards, providing

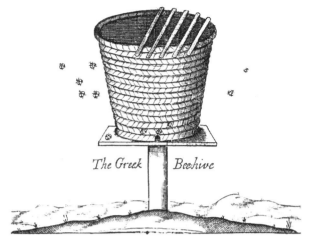

FIGURE 9.1 The Greek bar hive described by Wheler (1682).

the bees with a guide for comb-building. Thus, the Greek hive was born.

The Greek design inspired a number of innovative beekeepers. In 1683, a beekeeper known as "J. A." described in a London periodical a movable-frame hive based on the Greek design. J. A.'s box hive had frames whose sides were angled to the same degree as the Greek hive (figure 9.2). Unfortunately, the top of the frames fit too closely, and the bees attached comb to the box. As unique as this hive was, the limited circulation of the publication doomed J. A.'s innovation to obscurity (Herrod-Hempsall 1930).

FIGURE 9.2 J. A.'s frame with the tapering sides influenced by the Greek hive (Herrod-Hempsall 1930).

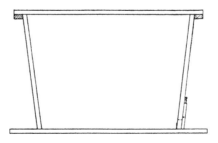

On the other side of the globe, Vietnamese beekeepers independently developed their own bar hive by modifying upright log hives. By placing strips of wood over the opening of the log hive in place of a board, they could lift the attached combs right out of the log.

In 1814, a Ukrainian beekeeper named Peter Prokopovich invented a hive that incorporated frames in part of the hive. Prokopovich's hive consisted of a box hive with three sections. The bottom two sections were not equipped with frames, but the upper section incorporated seven rectangular frames that slid into the hive from the sides. Holes in the bottom of each section permitted bees access to the various parts of the hive. Propokovich constructed his frames at a size that encouraged the bees to build combs on the frames, but the frames fit close to the sides of the hive and needed to be cut away for removal. Prokopovich published several papers in Russian beekeeping journals, and his hive was described in a French journal in 1841 (Crane 1999). But as with J. A., the small readership of the journals restricted Prokopovich's influence.

Francois Huber's leaf hive (figure 9.3), another early frame hive invented in the eighteenth century, was primarily intended as an observation hive. Its rectangular frame was divided into a narrow upper section and a larger lower section. The beekeeper placed comb in the upper half to give the bees the idea that they should build their comb within the limits of the frame. These frame sections were hinged together like the pages of a book. Frames with solid sides were placed on each end to make the sides of the hive, and a small opening was cut near the bottom edge of each frame to serve as an entrance. The whole package was then tied together. When Huber's assistant wanted to examine the frames (Huber was blind by age 15), all he needed to do was untie the frames and carefully spread them out.

The Bevan hive (figure 9.4) was a bar hive that incorporated the supering system of working bees. Seven bars were placed

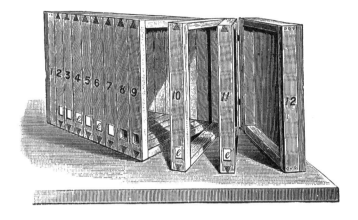

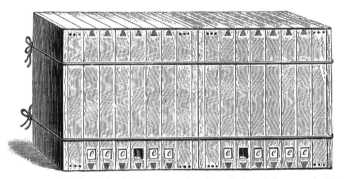

FIGURE 9.3 Huber's leaf hive, open (top) and closed (bottom). Note the string that is used to keep the frames closed, and the various entrances (*e*) that can be placed into the hives (Cheshire 1888).

across the top of a wooden box. As the box filled, additional barred boxes could be stacked on top, and the whole thing covered with a board containing holes for glass jars. Such an arrangement should have made the Bevan hive popular with beekeepers used to supering their skeps and working with glasses, but it never worked as promised. Because the sides were perpendicular to the base the bees attached the comb to the

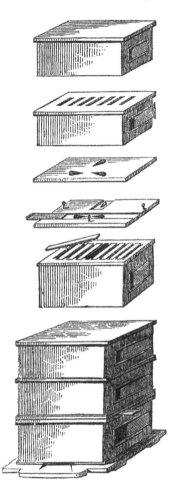

FIGURE 9.4 Bevan's bar hive (Taylor 1860).

sides, and the bars did not easily lift out. Moreover, wood in England was expensive compared to straw.

Movable-frame hives such as the Bevan hive, as well as the Munn and Debeauvois hives displayed at the Crystal Palace Exhibition, were not paid much attention, despite their novelty. In a contemporary description of the various hives at the exhibition, Richardson and Westwood (1852) classified these hives as "hives in which increased accommodation is not given to the bees," contrasting them with hives that use jars, ekes, collateral expansions, and the like. The *Illustrated London News* considered these hives with movable frames to be observatory hives and did not expect that they would influence the future of beekeeping.

The failure to see the advantages of these hives was in part due to the inherent difficulty of using them. Anyone who has worked honey bees knows that, if the bars or frames are too close together, the bees glue them down with bee glue, called propolis. If the bars or frames are too far apart, they build "burr comb" between the comb on the bars to bridge the space. Therefore, the Munn hive, the Debeauvois hive, and Golding's bar hive were not, in practice, easier to use than skeps or octagonal box hives. The concept missing from all of these hive designs was the "bee space." This critical dimension, about ⅜", is too narrow for the bees to build comb, and yet too far apart to glue together with propolis.

It was the American, Lorenzo Langstroth, who first incorporated this bee space in the design of his hives, though it was not recognized as a major discovery at the time. In his *History of American Beekeeping* (1938), Frank C. Pellet wrote, "The public failed to understand that the fundamental part of [Langstroth's] patent was the bee space." How such a major innovation of beekeeping could be overlooked is difficult to comprehend, considering that it changed apiculture in ways that continued to resonate over 150 years after the patent was filed. However, when the original patent is examined, the oversight comes as little surprise.

Langstroth started keeping bees with a log hive in 1838, but soon obtained two of Bevan's bar hives (Pellett 1938, Brown 1994) following consultation of references written by Bevan (1838) and Huber (1808). Huber (1808) extolled the value of the leaf hive, and wrote that to best entice the bees to "work the wax," the combs must "preserve the law which establishes an equal distance [between the combs] throughout the whole [hive]." In Bevan's hive, the bees constructed combs that hung down from a top bar and were connected to the sides of the hive. The difficulty in working this hive helped Langstroth realize that to have a truly useful hive, he must prevent the bees from attaching the comb to the hive's sides. Langstroth may have been familiar with Munn's hive design, in which the frames had ½" of clearance from the sides of the box and could be pulled out of the hive for examination (figure 9.5) (Crane 1999). However, Munn's frames were flush with the bottom of the hive; they were quickly glued to the base and became difficult to remove. Indeed, Alfred Neighbour is reported to have written, "This invention [Munn's hive] was of no avail to apiarists" (Cook 1884). It is surprising that skep beekeepers did not discover the bee space sooner. When bees

FIGURE 9.5 Munn's hive showing movable frames for examination (Cook 1884).

extend their comb inside a skep, they instinctively stop when the comb is within a bee space from the bottom. Filled skeps are easy to lift up because the comb is not glued or flush with the bottom.

The solution to the problem of movable frames came to Langstroth in the fall of 1851. His journal records the details: "If the slats [sides of the frame] are made so that [they] are about three-eighths of an inch from the sides of the hive, the whole comb may be taken out without at all disturbing it by cutting." He later wrote that the bottom of the frame should also be ⅜" from the bottom board (Naile 1976). With those words, Langstroth had described what we now call the "bee space." He proceeded to design a hive based on this space around the frames, but he had to wait until the following spring to test his new design.

Langstroth moved quickly. Henry Bourquin managed his apiary and applied his woodworking skills to make the needed changes to Langstroth's hives. By the end of the summer of 1852, Langstroth had over 100 movable-frame hives. He wrote in his private journal, "Imagine me so absorbed in manipulating these frames, with the bees upon them—removing from the hive and replacing them—shaking the bees from them, and changing their relative positions, etc." (Naile 1976).

On October 5, 1852, Langstroth received a patent for his hive, and it was numbered Letters Patent No. 9,300. A year later, he published *Langstroth on the hive and the honey-bee; a beekeeper's manual.* Langstroth (1853) did not illustrate his hive in the first edition of the book, and he did not divulge its dimensions (Crane 1999). The patent papers, however, included a diagram of the hive and more specific details.

Langstroth's patent claimed five inventions: a shallow chamber above the hive body, movable frames that did not require the cutting of the comb, a divider to control the size of the hive body, the use of double-paned glass as the sides of the

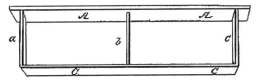

FIGURE 9.6 The illustration of the movable frame in Langstroth's original patent (Langstroth 1852).

hive, and a wax moth trap for the entrance. The crucial discovery, the bee space, is not mentioned by name, but is hidden in lines 79 through 89 on the first page:

> There should be about three eighths of an inch space between *a* and *c* [figure 9.6], and the sides and C C and the bottom board of the hive [*sic*] this will prevent the bees from attaching the frame to the sides or bottom board of the hive, hindering its easy removal, and will allow them to pass freely between the sides and the bottom board, and the frame so as to afford no lurking place for moths or worms.

This understated description of the most critical aspect of Langstroth's hive design contributed to the public failure to immediately recognize the true importance of his innovation. Indeed, the entire patent is written in a convoluted manner and is not easy to follow: figures are not presented in an orderly fashion, some labels are left off, and the changing orientation of the various figures makes the patent difficult to understand without careful study.

Figure 9.7 shows the hive from the front, depicting the cover closed over the hive and the shallow chamber above. The cover was hinged on the front so that it could be lifted from the back, revealing the hive inside (figure 9.8). From the back, the hive body can be seen within its glass box. Langstroth advocated double-paned glass for the inner hive because the dead-air space

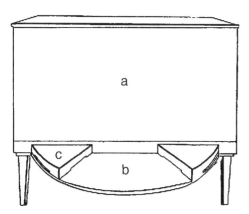

FIGURE 9.7 The front of Langstroth's patented hive (Langstroth 1852).

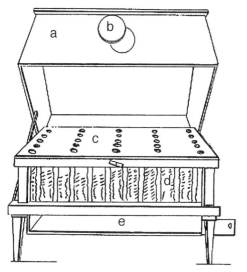

FIGURE 9.8 Langstroth's hive opened from the back, showing the glass-enclosed frames within (Langstroth 1852).

was an excellent way of insulating the hive. Moreover, the outer pane would be cooler than the inner pane, allowing any moisture in the hive to condense on the outer pane.

Separating the hive body from the shallow chamber above it was a board with 25 zinc-lined holes that were each ⅜" in diameter (figure 9.8). These openings permitted the bees to move from the hive body to the shallow chamber and back. The chamber could also be fitted with frames that incorporated the critical "bee space" between the frames and the sides of the hive. This, Langstroth claimed, would permit the apiarist to easily remove sealed honeycomb without disturbing the bees. The chamber could be filled with standard hive frames or small boxed frames for comb honey. In a link to the days of glass jar beekeeping, Langstroth (1859) illustrated his hive with a full set of glass tumblers and small bell jars (figure 9.9), although he did not generally recommend them.

FIGURE 9.9 A Langstroth hive with a full set of glasses and bell jars (Langstroth 1859).

The hive rested on legs that were 1½" longer in the back than in the front, giving the entire hive a slight forward slope. This would help drain rainwater off the top of the hive and away from the entrance, preventing it from flowing into the hive. The base also had a screened opening for ventilation, which could be closed off with blocks that were pushed in below the bottom and secured with a sliding panel.

The unique dimensions of Langstroth's hive have been the subject of much speculation. It has been suggested that Langstroth simply followed Bevan's dimensions for 18" by 18" bars, or he used a champagne box as the original hive (Crane 1999). Brown (1994) suggested that Langstroth simply cut the standard 18" by 12" glass pane in half to make the 18" by 6" glass sides for the sides of his shallow hive body. By 1853, Langstroth's hives were no longer made of glass, and the internal measurements, 18 5/16" by 14 5/16", were standardized for wood construction and to allow for wood shrinkage.

It is clear that Langstroth applied the bee space concept to his hive design, but he did not invent the term. In *A Practical Treatise on the Hive and the Honey-bee*, Langstroth (1859) never referred to "bee space," nor did he detail its significance. It is illustrated in his hive plans, but the reader would have had to use a ruler to measure the scale drawings in order to find that the space surrounding the frames was 3/8" to ½". This omission was probably due to Langstroth's fear of patent infringement. He wrote, "The reader will bear in mind, that those only who have purchased the patent right—Ministers of the Gospel excepted—can legally use these hives."

The term "bee space" seems to have been introduced by James Heddon in 1885. On page 78 of his pamphlet "Success in bee-culture," Heddon attributed the idea to Langstroth: "The under-side of our Honey Board has an even surface, and when placed on the hive, rests bee space [*sic*] above the tops of the movable frames, because they rest 3/8 of an inch below the top edge of the hive. This is one of Father Langstroth's inventions, and in my estimation best of them all,… To better describe its value, I will quote from the specifications of his patent papers, long since expired…" He then printed several paragraphs from Langstroth's patent papers, none of which use the term.

Heddon (1885) also provided a more formal definition on page 123. "The term 'bee-space,' does not only denote a space that will admit of the passage of a bee, but it refers to that space in which bees are least inclined to build brace-combs or place propolis, or bee-glue; which is a scant ⅜ of an inch."

It did not take long for this term to catch on in beekeeping literature. In the 1887 edition of *The ABC of Bee Culture*, Root added Heddon's bee-space definition to the glossary. The next year, Frank Cheshire (1888) used the term in his British book and included labeled illustrations of the bee space inside straw skep hives (figure 9.10), and in Langstroth's (figure 9.11) and Heddon's hives. In 1889, Langstroth formally took ownership of the term with Dadant's revision of *Langstroth on the Hive and the Honey Bee*. Dadant (1889) wrote, "Mr. Langstroth invented the top-opening movable-frame hive, now used the world over, in which

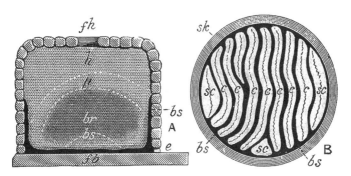

A, Vertical Section—*fb*, Floor Board ; *e*, Entrance ; *br*, Brood ; *p*, Pollen ; *h*, Honey ; *fh*, Feeding Hole ; *bs*, *bs*, Bee-space. B, Horizontal Section—*sk*, Skep-Side ; *c*, *c*, Combs ; *sc*, *sc*, Store Combs ; *bs*, *bs*, Bee-space.

FIGURE 9.10 The bee space inside a straw skep hive (Cheshire 1888).

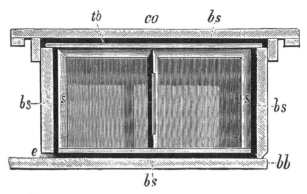

co, Cover; bb, Bottom Board; and e, Entrance of Hive; bs, bs, Bee-space; tb, Top Bar; and s, s, Sides of Frames.

FIGURE 9.11 The first time the bee space was labeled inside a Langstroth hive (Cheshire 1888).

the combs are attached to movable frames so suspended in the hives as to touch neither the top, bottom, nor sides; leaving between the frames and the hive walls, a space of one-fourth to three-eighths of an inch, called bee-space."

Langstroth's concern about the theft of his intellectual property turned out to be well founded. In 1867, H. A. King licensed Langstroth's hive design and paid the inventor for three years. In 1870, King terminated his agreement with Langstroth, citing numerous modifications he had incorporated into the hive and claiming that he was no longer selling hives based on Langstroth's design, and yet continued to produce hives using the critical bee space. Langstroth sued King, and King, in preparations for his defense, traveled to Europe to find evidence that Langstroth was not the inventor of the movable-frame hive. Indeed, upon his return King cited Munn's and Debeauvois' hives as evidence that Langstroth was not the first to use frames, and argued that the patent should therefore be voided (Pellett 1938). In reality, all this proved was that King did not understand the significance of the bee space. Whether or not a hive contained movable frames was not the issue; it was the space between the frames and the hive boxes that was critical. The case was never decided, and Langstroth decided to let users of his invention pay him or not: their conscience would be their guide.

Langstroth's application of the bee space in his early hive design was his greatest contribution to beekeeping. Unfortunately, his failure to specifically name the space between the frames and the hive, combined with his concern about patent infringements, contributed to the public's misunderstanding of the most important feature of his patent. In reality, Langstroth did not invent the bee space—the bees did. But he was the first to discover its importance, and incorporate it into the design of his hive. In spite of the intellectual property theft, the bee space proved to be the most important beekeeping discovery made in his time, and its influence continues to this day.

CHAPTER 10

Resistance to Change

THE RESPONSE TO LANGSTROTH'S NEW HIVE WAS low-key, to say the least. In America, many beekeepers continued to use log hives or bee gums for decades. But by the 1870s, most American beekeepers who were purchasing hives or hive plans were taking advantage of the concept of bee space.

In England, Langstroth's discovery was first mentioned in a paper by T. Woodbury (1862), but it is apparent that the true nature of the bee space was not yet fully appreciated. Woodbury wrote, of Langstroth's frame hives, "In these hives every comb is constructed in a frame, and is not allowed to touch any part of the box. They have, therefore, the advantage over all others, that every comb may be extracted, examined, and replaced without injury, or danger to the life of a single bee." He cited increased costs as the major disadvantage. Although Woodbury did not give the dimensions of the bee space, he noted that the frames were fitted in such a way as to allow "the bees free passage above and below, as well as at the sides."

Like their American counterparts, English beekeepers also failed to grasp the significance of the bee space. Instead of a beekeeping revolution taking place, battle lines were drawn over straw versus wood, and skeps versus boxes.

The impact of movable-frame hives on British beekeeping can be seen in subsequent editions of popular beekeeping books, such as Alfred Neighbour's *The Apiary* (1866, 1878). Neighbour was a major purveyor of beekeeping supplies, and his books document the demands of the English market. Neighbour described sixteen hives in his 1866 edition and nineteen in the 1878 edition. Eleven hives appear in both editions, including Nutt's collateral hive, the Stewarton hive, the Improved Cottage hive, and of course Woodbury's hive based on Langstroth's work. New to the 1878 edition were several other hives with movable frames, such as the Cheshire hive and the Abbott frame hive. Surprisingly, Neighbour also introduced several new straw skeps in the 1878 edition, including a new common cottage hive (figure 10.1). These straw hives, much cheaper than the new movable frame hives, illustrated the continued popularity of the skep sixteen years after the Langstroth hive was first mentioned by Woodbury.

In the first edition of his popular *British Bee–keepers Guide Book* (1881), Thomas Cowan wrote, "The straw skep of our forefathers, together with the ignorance and superstition connected with it, is steadily dying out." Unlike Neighbour, whose book had appeared just three years earlier, Cowan included not a single skep among his hive descriptions. Frank Cheshire

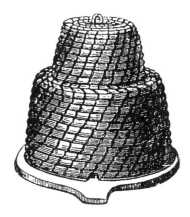

FIGURE 10.1 Common cottage hive sold by Neighbour (1878).

followed suit in his 1888 book, signaling, perhaps, that the last days of skep beekeeping were approaching. Accordingly, a reader might expect that subsequent editions of Cowan's book would document the further decline of the skep. In fact, just the opposite is true. The twenty-first edition of Cowan's book, published in 1913, includes photographs of skeps as part of the beekeeper's inventory of tools. Although wooden boxes with movable frames were replacing skeps as the hive of modern beekeepers, skeps remained popular for swarm capture.

In 1913, straw skeps were still inexpensive and their light weight made them easy to carry on bicycles (figure 10.2), a common mode of transportation at the turn of the century. The practice of using skeps for swarm capture evolved out of the skep

FIGURE 10.2 Skeps being transported on a bicycle (Cowan 1913).

beekeeper's practice of driving bees: encouraging bees to leave a skep and move into another one. Victorian skep beekeepers did not kill their bees over a brimstone pit as was common a century earlier, but managed their bees by driving them into a new skep. They also used skeps to collect swarms and to merge weak hives into stronger ones. The old-fashioned skep still had important uses.

The battle over hive materials was also heating up during this time. "Straw hives, well sewed with split canes or bramble briers, are incomparably better for bees than any other kind of hive yet introduced. Nothing better is needed, and we believe nothing better will ever be found out," wrote A. Pettigrew (1870), one of the last holdouts for straw hives in the late nineteenth century. In his 1870 book, *The Handy Book of Bees*, he went on to write, "Where straw hives cannot be obtained, wooden boxes are used; but they are very objectionable in every sense, save, perhaps durability. Hives made of wood, at certain seasons condense the moisture arising from the bees, and this condensed moisture invariably rots the combs." Such strong feelings were held by a number of British and European beekeepers. *The British Bee Journal* carried short articles and questions from straw-hive beekeepers well over fifty years after Langstroth's hive was introduced. This is not to say that skep beekeepers did not see the advantages of movable frame hives. On the contrary, many adopted this new technology, but did so with straw hives.

The merging of frames with straw required a departure from earlier bar skeps, which had incorporated fixed wooden slats to support comb. Making a frame fit a traditional basket hive was challenging, but in England several beekeepers were successful at making boxes with straw sides and incorporating hanging frames in the boxes. In Germany, a different approach was used. Thus, two fundamental designs emerged from beekeepers' desire to adopt movable frames for straw hives.

One popular such hive was developed by T. Woodbury. Langstroth had presented Woodbury with a copy of his book,

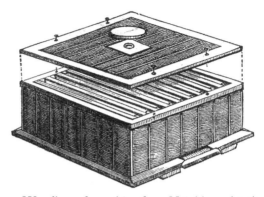

FIGURE 10.3 Woodbury frame hive from Neighbour (1878).

which Woodbury used to develop the first hanging frame hive in England (IBRA 1979). The "Woodbury Hive," as it was popularly known, was a wooden hive with Langstroth-style frames. Woodbury later introduced a hive body that used straw instead of wood, recommending that the stock be kept in such hives because of straw's superior ventilation and moisture control, which kept the colony healthier (Neighbour 1878).

The Woodbury straw frame hive (figure 10.3) was 14½" square and 9" deep. It had a 1"-thick wooden frame at the top, which held 10 hanging frames that were 13" by 7¼" in size. These hanging frames were held a bee space apart by notches along the side rails. The cover was also made of straw, with a small hole in the center to permit the use of a glass super or another square box (Neighbour 1878).

George Neighbour and Sons later modified the Woodbury straw frame hive with the aid of a machine that formed straw sections. "Neighbour's New Straw Frame Hive," as it was billed, had wooden corners and edges that formed a scaffolding for the straw sides (figure 10.4). Neighbour's addition was the small glass windows that permitted the beekeeper to view the interior without disturbing the bees, a feature he also included in his elaborate skep hives. These hives used a framed straw "crown-board" with

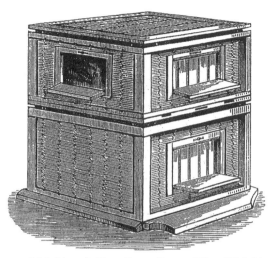

FIGURE 10.4 Neighbour's New Straw Frame Hive (Neighbour 1878).

a small hole for spring feeding and zinc sliders that opened the hive body to the supers. These openings incorporated queen excluders to prevent the queen from laying eggs in the supers. Queen excluders, first used with skeps in the early nineteenth century, were wooden grates with slats spaced widely enough to permit worker bees to pass through, but narrowly enough to block the queen. With later hives they were made of zinc with perforations of the appropriate size. Neighbour recommended that straw honey supers with three windows be placed upon the single-windowed hive body. These hives were lightweight, clearly facilitated ventilation, and were of the highest craftsmanship of any of the late nineteenth-century hives. It was expected that they would be kept inside a wooden box cover that would protect the straw hive and decorate the garden (Neighbour 1878).

A third example of a straw hive with frames was the Sherrington hive (figure 10.5). This gabled-roof hive was made by resting a square hive body on a wooden bottom board, which was furnished with pegs to hold the hive in place. Inside were eight frames, apparently without bottom rails. Upon the straw

FIGURE 10.5 The Sherrington hive with its gabled cover (Hunter 1875).

box rested the highly gabled roof, which was placed over glass jar supers (Hunter 1875).

German beekeepers incorporated the benefits of straw with frame hives in a very different system. The "Bogenstülper" hive (figure 10.6) was invented by C.J.H. Gravenhorst in 1865. The name Bogenstülper comes from the German words for bowed

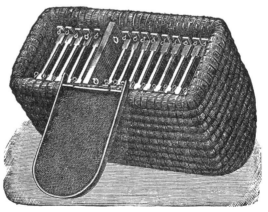

FIGURE 10.6 The Gravenhorst hive or "Bogenstülper" (Dadant 1889).

FIGURE 10.7 The "Bogenstülper" was a large hive that was worked from the bottom. Note its large size in relation to the beekeeper (Dadant 1889).

or arch (bogen) and tilted (stülper) and refers to how the hive was to be used (Herrod-Hempsall 1930). This large straw hive, shaped like an American mailbox, was designed to be worked from the open bottom, as were traditional skep hives. Inside were 16 frames held in place by pins at the bottom (figure 10.7). In order to work the hive, the beekeeper would lift the hive off the shelf where it was kept and then turn it over onto its top. Some German beekeepers used a special cart designed to hold the hive and support it when inverted. Then the pins of the frame to be examined, as well as those on either side, were removed. This permitted the frames to be moved slightly, freeing them from the sides of the hive and allowing for easy removal of the center frame. Dadant (1889) included a brief discussion of these hives in "Langstroth on the honey-bee," in which he wrote that this hive was quite heavy when filled with honey.

These straw frame hives were brought into the marketplace because beekeepers demanded the benefits of straw. They eventually fell out of favor, however, as people discovered that they were not actually cheaper than box hives in the long run. One benefit of straw skeps was their low cost, but the new framed straw hives, with their machined sides and wooden frames, were more expensive, and were just as labor-intensive to maintain as the traditional skeps. If a beekeeper did not take care to protect these hives with wooden covers, they would suffer the same deterioration as skeps when exposed to the elements. Moreover, the lighter material could not support the weight of the increased yields the new frame hives could produce, and they needed constant repairs or, worse, replacement (Herrod-Hempsall 1930).

Decades of skep beekeeping in England resulted in beekeepers skilled in the manipulation of swarms, and this skill found a niche in the new techniques of movable-frame hives. Many box hive beekeepers procured new colonies cheaply by purchasing bees that older, more conservative skep beekeepers were going to kill as part of the honey harvest. More commonly, new colonies were obtained by using skeps to capture bivouacking swarms. It was an easy process to hold an empty skep under a branch laden with newly swarmed bees and shake them into it (figure 10.8). The skep could then be placed in a burlap bag and carried to the site of the new apiary. Skeps were a popular and inexpensive way of transporting bees over long distances. Cowan (1913) illustrated a skep carrier that was used for transport of bees on trains (figure 10.9).

Skep use has continued into the twenty-first century. E.L.B. James (1950), author of a popular British beekeeping book, mentioned that driving bees from one skep to another was still being practiced in 1947. Skeps with bees were available for purchase in Spain and the Netherlands in the 1990s, and I have observed skeps being used for swarm capture at

FIGURE 10.8 Using a skep to capture a swarm (from a lantern slide in the Kritsky collection).

FIGURE 10.9 A skep prepared for train transport (Cowan 1913).

Pluscarden Abbey near Elgin in Scotland as recently as 2002. Beekeepers in Europe can purchase skeps for swarm capture today over the Internet. Despite over a century of hive design innovation, the relatively inexpensive lightweight straw skep still finds a place in apiculture alongside newer, more modern hive designs.

CHAPTER 11

Alfred Neighbour: British Beekeeping Ambassador

W HEN WE CONSIDER THE HISTORY OF beekeeping, we often think of those individuals who invented and designed new hives. And indeed, the 25 years following the Crystal Palace Exhibition and Langstroth's discovery of bee space witnessed an explosion in hive design, especially in England. But it was Alfred Neighbour, a purveyor of beekeeping supplies, who through his savvy marketing and good contacts was responsible for a fundamental change in English beekeeping.

Alfred Neighbour (figure 11.1), born in 1825, was the son of George Neighbour, who, in 1814, started selling "bee furniture," the nineteenth-century term used to describe hives and bee houses. The year before Alfred was born, Thomas Nutt, the inventor of the Nutt Collateral Hive, gave George the rights to sell this new hive, and thus began a long association between the two men. Nutt had a custom of visiting people who had purchased his collateral hive to see how it was working for them. Accompanying him on many of these tours was the young Alfred

FIGURE 11.1 Mr. Alfred Neighbour (Cowan and Carr 1891).

Neighbour, and this early mentoring gave Alfred a different perspective on beekeeping beyond simply managing hives. Nutt was deeply involved in the science of apiculture, and believed that regulation of temperature would help control the bees. He was also a tireless self-promoter, and his book, *Humanity to Honeybees*, was essentially an advertisement for his philosophy of beekeeping (Cowan and Carr 1891).

Alfred Neighbour eventually joined his father as a partner in George Neighbour and Sons, selling beekeeping furniture. Following Nutt's example of self-promotion, they marketed themselves with a public apiary at the London Zoo (figure 11.2). The bee house was emblazoned with a sign that read: "BEES Presented by G. NEIGHBOUR AND SONS." Originally, only Nutt Collateral Hives were featured in this display, but these were later replaced by five different types of hive, including an

FIGURE 11.2 The exterior of Neighbour's public apiary at the London Zoo (Neighbour 1866).

observatory hive, a crystal observatory hive, an improved cottage hive with five glass supers, and two box hives (figure 11.3). All of these hives were available for purchase at the Neighbours' store. The apiary was a particular favorite of the royal princes and princesses. One morning, Alfred was working the bees when the royal children visited, and he showed them the queen bee and explained the basics of beekeeping. The Neighbour name achieved national fame with their display at the 1851 Crystal Palace Exhibition, and they expanded their operations to two London storefronts the following year.

Alfred was a close associate of Henry Taylor (author of *The Bee-Keepers' Manual*), and it was Taylor who told Alfred that Thomas Woodbury had been offered a "Ligurian" queen bee. Ligurian bees, found in Italy, were favored for their productivity and docile nature. Alfred, ever the businessman, wrote to Woodbury's source and requested a queen for himself. The two

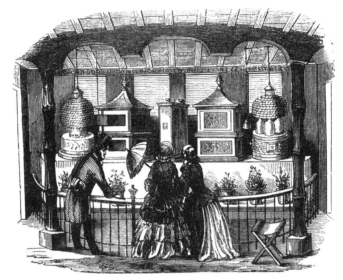

FIGURE 11.3 The interior of Neighbour's London Zoo apiary (Neighbour 1866).

queen bees arrived on July 19, 1859, and soon changed British beekeeping by becoming the stock for a new, more productive race of bees working English hives. The event also sparked a close friendship between Woodbury and Alfred Neighbour.

The second Crystal Palace Exhibition in 1862 provided Alfred with another chance to showcase new hives, as well as promote the Ligurian race of bees. His display included an observation hive, a glass-enclosed hive that permitted people to watch the bees at work, filled with the new Ligurian honey bees (figure 11.4). Also on display was the new Woodbury hive (figure 10.3). In an article for the *Gardeners' Weekly* magazine, S. Hibberd wrote that the Woodbury hive was "a novel construction, combining all the best features of the best bar boxes, and adding some new ones of great value and importance. We recommend every bee-keeper to become possessed of this admirable contrivance, with which Mr. Woodbury has accomplished

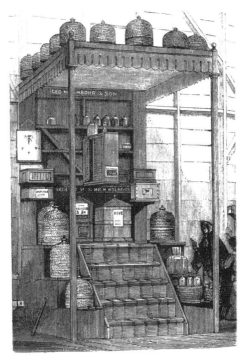

FIGURE 11.4 Neighbour's display at the 1862 Crystal Palace Exhibition (Neighbour 1866).

wonderful things in the multiplication of the new race of Ligurian bees." This public introduction of both the Ligurian or "Italian Alp" bee and the movable-frame hive based on the bee space brought considerable attention to the Neighbour firm (Neighbour 1866).

The following year, Alfred put on a large display at the Bath and West of England Agricultural Show. Again he featured the Ligurian bee and Woodbury's hive. He also exhibited his new wax foundation, which the *Journal of Horticulture* (Neighbour 1866) reported "as being well worth examination." Artificial comb, now called wax foundation, is a sheet of beeswax that has been run through a press to produce the outline of bee-cells over its surface. It encourages the bees to produce more even combs with fewer drone cells, yielding more honey. It was invented in Germany in 1843, and its use spread throughout Europe during the 1850s (Crane 1999). The correspondent for the *Journal of Horticulture* commented, "It is almost unnecessary to state that this unique and instructive stall was crowded throughout the week, and we hope its financial results were such as will lead Messrs. Neighbour to continue their attendance at the Society's meetings"—a testament to the reputation that George Neighbour and Sons enjoyed as a result of their exhibitions.

The next World's Fair was held across the Atlantic, in Philadelphia, in 1876. The great Centennial celebration of the United States included displays from around the world, and George Neighbour and Sons were among the British exhibitors in the Agricultural Hall. Their relatively small setup was across the aisle from the French wine exhibit and in front of a display of fine English chocolates and sauces. There is no formal inventory of all the goods that they brought; however, a photograph of the hall's north wing shows the sign "Geo. Neighbour and Sons. Beehives" and part of their display. On the table were at least ten hives, including several varieties of straw skeps, such as the Crystal Palace skep hive and the Improved Cottage Hive (figure 11.5).

American reactions to the Neighbour display was tepid, demonstrating Americans' lack of enthusiasm for the straw hives preferred by the Brits. The Burr and Burr (1877) Memorial of the International Exhibition described Neighbour's hives as having "the form of small domes, and…made of ropes of straw." The December 1876 (Anonymous 1876) edition of the *American Bee Journal* provided a detailed description of the Neighbour display:

> The largest display of apiarian supplies was that of Messrs. George Neighbour & Sons, 149 Regent St., London, England.

FIGURE 11.5 Neighbour's beehive display at the 1876 Centennial Exhibition in Philadelphia (Kritsky collection).

It comprised their cottage hive, observatory hive, cottage frame-hive, divisional super, sectional boxes, feeders, wax guides and plates for making them.

An examination of these was very interesting to one familiar with our American inventions. The "Cottage hive" is of rustic appearance, and neatly made of straw, strengthened with hoops, fitting closely to the wood. It is a two-story observatory hive. It has three windows in the lower story, with a thermometer to indicate the temperature; showing the bee-keeper when to open the three entrances to the upper story, over which there are three large bell glasses to be filled with surplus. The upper story fits over these glasses and may readily be removed for inspection. The bottom board is hinged to the lower story.

The "frame hive" has moveable frames fitted with staples to keep them at regular distances, resting on a zinc ledge above.

The "frame unicomb hive" is a novelty which must be seen to be appreciated. It is constructed with glass sides (for observation) and protected with Venetian blinds.

The "divisional super" is very much the same as our sectional boxes. It contains 7 sections or frames; the entrance being through perforations in a sheet of zinc, large enough to admit workers, but not the queen or drones.

The American attendees were not impressed. The author of the above description was polite, but this was not the case when Neighbour exhibited in Paris in 1878. An American reviewer (Argus 1878) wrote, "Although the display is large, exhausting, I think their entire catalogue, [the display] was in many respects a duplicate of their Centennial exhibit, and contained little, if anything, that would interest American bee-keepers enough to adopt." Beekeeping preferences remained fundamentally regional, and the straw hive simply did not catch on with Americans.

From Alfred's point of view, however, Neighbour's Centennial exhibit was far from a failure. It was awarded a prize medal (figure 11.6) with a citation reading, "For a large and varied collection of economical beehives so arranged that the honey can be

FIGURE 11.6 The type of medal awarded to Neighbour for his Centennial Exhibition display. Photograph by the author.

taken without the destruction of the bees. Special attention is directed to the Unicomb Hives with Venetian blinds to allow the bees to be exposed to light, whilst the sun's rays are excluded. Also to a Honey Extractor by centrifugal force, which removes the honey from the combs without injuring the latter, which can be returned to the hives." Honey extractors, which I will discuss in chapter 16, were relatively new at the time of the Philadelphia Exhibition. Neighbour wrote to the *American Bee Journal* on February 2, 1877, informing the *Journal* of his medal and the reasons for receiving it. No doubt, he had hoped to generate some business from the award, but straw skeps, no matter how complex, never caught the attention of the American market. Neighbour's British invasion did not succeed, but he continued to exert an influence on British beekeeping until his death in 1890.

CHAPTER 12

Transition

B EEKEEPING AROUND 1900 WAS IN A STATE OF transition. In all of the editions of his *British Bee-Keepers Guide Book* from 1881 until 1913, Cowan claimed that the skep hive was "steadily dying out" and being replaced by box hives with movable frames. However, during the same period, the *British Bee Journal* attested to the continued popularity of skeps and recorded an extensive debate on what kind of frame hive to use. These books and the *BBJ* do not, however, report on contemporary practice in the apiary of the cottage beekeeper. A photograph taken at that time provides an idea of the kinds of practices that were going on during this transitional moment.

The photograph (figure 12.1) was taken by Francis Poppleton of the Exchange Studio in Doncaster, southeast of Leeds. The beekeeper was a level-crossing keeper, an attendant who opened and closed the gate when trains crossed the road, and who controlled the signal tower that informed oncoming trains whether the track was clear. He and his family lived in the brick house (seen in the background of the photograph) that adjoined the level crossing. The specific location is along the London to Edinburgh line, on which there are 18 such crossings. The level-crossing keeper was required to be present at specific times, but this left open long stretches of time during the day. Beekeeping

FIGURE 12.1 Photograph of an English beekeeper taken between 1880 and 1900. The beekeeper's identity is unknown, but he is not Richard Frow, a well-known beekeeper and stationmaster at Wickenby (Kritsky collection).

was an ideal second job, and the photograph shows the subject proudly standing in his apiary.

Our beekeeper was managing at least 14 hives at the time the photograph was taken: eight wooden box hives, two straw skeps, and four cylindrical hives of some kind. Studying the hives in this photograph in detail and comparing them to contemporary publications offers a brief glimpse into the life and work of a late nineteenth-century English beekeeper.

The beekeeper's two straw skep hives were placed on wooden stands that raised them about 10 inches off the ground, and they were covered with milk pans held down with a large hewn stone (figure 12.2). These skeps were likely used for swarm capture, and their presence in the apiary with a pan cover suggests that the skeps were in full use at the time.

FIGURE 12.2 One of the
two skeps in the apiary
(Kritsky collection).

The level-crossing apiary had four unusual cylindrical hives (figure 12.3). These were composed of one or two cylinders, with the top cylinder resting within the wider bottom one. A quilted cloth covered the top section, which, like the skeps, was further protected with a milk pan. Based on a review of the literature, these hives were probably made from wooden cheese boxes. In his historical review of beekeeping (1930), Herrod-Hempsall included a photograph of a cheese box hive and wrote, "In the days of our youth, when short of hives, we frequently put swarms and driven bees into cheese boxes" (figure 12.4). We can infer that our beekeeper was likely short of hives when the photograph was taken.

The eight remaining hives are the more elaborate wooden cottage hives with movable frames (figure 12.5). There were many purveyors of box hives in England at the time, and these hives have the external features of several popular models. Regardless of the exact design, the hives' external features illustrate how English hives differed from their American counterparts.

FIGURE 12.3 A hive fashioned out of cheese boxes (Kritsky collection).

FIGURE 12.4 A cheese-box hive (Herrod-Hempsall 1930).

FIGURE 12.5 A framed cottage hive used by the level-crossing keeper (Kritsky collection).

These differences would help the bees survive the typical English winter, rainier and milder than winter in many parts of America.

The deep gabled roof is reminiscent of the Abbott Standard Hive first introduced in 1873. The sloping roof allowed water to more easily flow off the hive, and the hole just below the apex provided ventilation. The ledge running around the base of the roof—the "plinth"—was nailed on to help the beekeeper lift the roof. Plinths were common features of British hives; they helped protect the seam that resulted from placing the roof or boxes upon other boxes. When the hive was prepared for winter, a thick quilt or mat of straw would be placed inside, upon the brood chamber, to help insulate the bees.

The lower box housed the brood chamber. A close examination of the photograph shows that our beekeeper was using two sizes of hive covers, which may have been the result of his using larger hives in "combination." Combination hives permitted the beekeeper to use 10 large frames perpendicular to the entrance, or 15 frames running parallel. When properly managed, this resulted in the production of sealed honey in the brood box. A section rack for the production of comb honey could be added above the honey super, which was placed above the brood box.

The third box hive from the left in figure 12.1 shows a shallow brood cover with a "lift" above it. Within this section a shallow honey super or another section rack could be kept. Many cottage beekeepers produced comb honey because its purity was guaranteed, and it did not need to be expensively extracted.

Five of the box hives have extended alighting boards called "swarming boards," which facilitated the transfer of swarms into box hives. The entrances of all the box hives had slightly sloping porches that protected them from rain.

Our beekeeper appears to be typical of late nineteenth-century English beekeepers. He honed his skills working the skep hive, was frugal in his choice of equipment (as evidenced by his cheese box hives), and was gradually transitioning to movable-frame hives. Smartly dressed in his railway uniform, he looks out at us with an expression of pride. Like many beekeepers today, he led a dual life and, fortunately for us, preserved features of his occupations in this photographic time capsule.

The American transition to modern beekeeping followed a different path. American beekeepers did not have skep or octagonal hives to compete with the Langstroth model. Instead, they used a wide range of hive sizes containing different types of frames.

By the time Langstroth (1859) published his book, *A practical treatise on the hive and the honey-bee*, he had already changed the design of his patented 1852 hive. His new hive had a portico-style entrance, and looked very different from the hives in use today (figures 12.6 and 12.7).

FIGURE 12.6 A beekeeper standing next to his apiary composed of Langstroth hives.

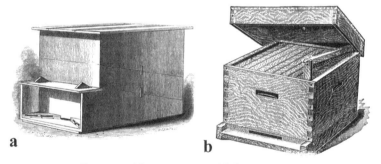

a   b

FIGURE 12.7 Langstroth's portico hive (a) (Langstroth 1859), compared to the "modern" hive (b) (Root and Root 1923).

Langstroth was not the only American patenting beehives. The U.S. Patent Office index records 558 patents for beehives between 1790 and 1873 (Leggett 1874). Many of these "inventions" involved only minor differences from other patented designs, which led to legal threats and damaged reputations. A few, however were based on different philosophies of beekeeping, in particular regarding frame construction and the design of the hive box.

An examination of late nineteenth-century beekeeping literature shows that nine major frame dimensions were in use in the United States (figure 12.8) (Root and Root 1913). In general, these frame dimensions were variations on two fundamental shapes: square frames versus long, rectangular frames. The proponents of the square frames observed that, as bees formed spherical clusters in the winter months, the square frames were more conducive to the bees' natural behavior.

FIGURE 12.8 The various frame sizes used in the United States in the late nineteenth century (Root and Root 1913).

Those who preferred the rectangular frame took a structural view, arguing that the low, wide boxes in which they were kept were more stable when stacked. Hives designed to hold square frames had a smaller base and, when stacked, were more likely to topple over. The rectangular frames had other benefits as well: easier uncapping of sealed honey, a better fit in the newly invented extractors, and less likelihood of splitting under the weight of the honey (Root and Root 1913).

To complicate matters, rectangular frame users were further divided into two camps: those who favored deep frames based on Langstroth's dimensions, and those who preferred shallower frames like those promoted by James Heddon (Pellett 1938). Heddon (1885) felt that Langstroth's frame was too big. He favored splitting the brood chamber into two shallow chambers, keeping a horizontal bee space through the middle of the brood. This permitted the bees to move about more easily when in their winter cluster. Moreover, Heddon could manage his hive by moving these smaller brood boxes rather than larger individual frames, and this sped up managing the hives and harvesting the honey (Root and Root 1913 and Pellett 1938).

Advocates for the larger frame included Charles Dadant, who credited Quinby with the large-frame hive. Dadant believed that honey yields were increased by multiplying the number of bees in the hive, and that smaller frames limited the queen's egg-laying opportunities (Pellett 1938).

Also debated was how the frames should be situated within the hives. Langstroth's design had a hanging frame with the top bar larger than the bottom, permitting it to hang within the hive (figure 12.9a). The thickness of the top bar and a wire guide bisecting the bottom of the hive box helped to accurately space the frames and lessen the production of burr comb. Others had different ideas. Quinby used two frame designs: one was similar to Langstroth's, and the other used a closed-end frame with wide side rails that were taller than the frame. This allowed the frame to stand on its side rails, which, when in

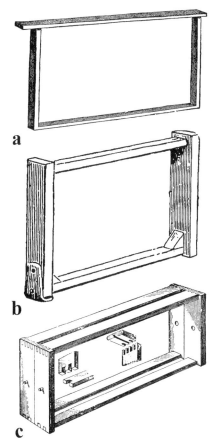

FIGURE 12.9 Comparison of competing frames in use by 1890: (a) Langstroth frame, (b) Quinby frame, which was held in place by a metal guide on the bottom, and (c) Danzenbaker frame with its pivoting support (Dadant 1889 and Root and Root 1913).

a

b

c

contact with each other, properly spaced the frames (figure 12.9b). Heddon and Danzenbaker recommended other styles of closed-end frames. Unlike Quinby's design, their side rails were flush with the top and bottom. Danzenbaker's design (figure 12.9c) included a pivoting pin on the side of the frame, which rested on cleats fastened to the inner wall of the hive. Heddon rested his frames on a thin metal ledge along the bottom of the hive box, and they were then compressed with wooden thumb-screws fixed to the sides of the hive boxes (figure 12.10) (Root and Root 1913).

FIGURE 12.10 Heddon rested his frames on a metal ledge that were held in place on the bottom of the hive by thumbscrews (Heddon 1885).

During these "frame wars" of the late nineteenth century, the greatest concern for American beekeepers was cost. Whatever the shape of the frame, smaller hives required less material, were less expensive, and were thus initially more popular (Pellett 1938). However, Dadant turned out to be correct with regard to frame size. Bees in the smaller hives filled the lower chambers mostly with brood—a fact that many small-scale beekeepers failed to notice when they harvested their supers. This also resulted in high levels of winter kill. Even Heddon himself suffered low honey yields and his hives fell out of favor, being rarely used by 1915 (Root and Root 1913 and Pellett 1938). The large rectangular frames, higher-yielding than the smaller rectangular frames and more stable than the square ones, eventually proved their superiority.

While frame shape was evolving, hive box design was undergoing changes as well. Langstroth's original portico hive

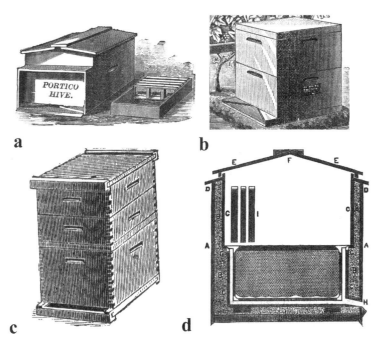

FIGURE 12.11 The major American hives in 1890 included the portico hive (a), the Simplicity hive (b), the dovetailed hive (c), and the chaff hive (d) (Root 1890).

incorporated the entrance and bottom in the hive body, and the sides were affixed directly to each other (figure 12.11). The 1890 *Bees and Honey Illustrated Catalog*, distributed by A.I. Root, sold four hives that became popular nearly forty years after Langstroth developed his first hive (Root 1890). The portico hive (figure 12.11a) was essentially Langstroth's original design. The Simplicity hive (figure 12.11b) used 10 Langstroth frames, but had a separate bottom board and a gabled or flat cover made of boards with cleated ends. The dovetailed hives (figure 12.11c) also had separate bottom boards and covers, but their sides were dovetailed together. The dovetailed hives also included a bee space above the eight or ten Langstroth frames, with the bee space below the frames created by the side rails of the bottom board. The cover was flat and cleated by a strip of wood on each end to prevent warping.

Finally, chaff hives (Root 1890) were double-walled hives that could be filled with wheat or oat chaff to provide insulation (figure 12.11d). These hives included tar paper on the bottom supports and a tin-covered gabled roof, both of which kept moisture from getting into the hive. These hives were popular in northern climates and were purported to reduce winter kill to only 3%. These "insulated" hives required less feeding than single-walled hives, and the bees required less honey to overwinter (Root and Root 1913).

As with the frame debates, economics played a major role in hive box choice. The Simplicity hives, with their interchangeable parts, were considerably cheaper than the other options. Beekeepers appreciated the dovetailed construction, which was more rugged than nailed butt-joint construction. Although slightly more expensive upfront, the dovetailed construction paid off over time, and within a few years most hives were built with dovetailed joinery.

The hive cover was also modified in the late nineteenth century. In Langstroth's original design, the hive was housed within a separate box with a flat roof. This flat cover was nailed to the box, and the entire upper box hinged open (figure 9.8) (Langstroth 1859). By 1888, Langstroth and Dadant were recommending a half-story cover that telescoped over the super (figure 12.7b) (Dadant 1889). By 1915, shallower telescoping covers lined with metal were available (Root and Root 1913). These, along with the flat cover with cleated ends (figure 12.11c), were widely adopted.

The last major change to frame sizes came about in 1920. This was the modified Dadant hive, which used 11 frames that were $2\frac{1}{8}''$ deeper than the Langstroth frame (Pellett 1938). This design, although smaller than Dadant's "Jumbo" frame, proved extremely popular because the increased frame size provided more space for winter honey stores.

All the parts of the modern hive were now on the scene. If today's beekeeper were to travel back to 1920s America, he or she would see familiar hive furniture that had survived the rigors of use and modification. Langstroth's original hive, with its portico entry and hinged lid, had evolved into dovetailed boxes resting on a separate bottom board and covered with a shallow box lid or cleated board. The modern hive had arrived.

CHAPTER 13

Bee Houses

THE BEE HOUSE OR HOUSE APIARY, A BUILDING designed to hold the hives for an entire apiary, was considered by some nineteenth-century beekeepers to be essential for modern beekeeping. Some were simple square buildings, while others were ornate octagonal constructions designed as much to enhance a garden as to convenience the beekeeper. As straw skeps became more elaborate and expensive, keeping them within a bee house or some other protective structure made sense, and nineteenth-century beekeeping literature documents many examples of bee houses in England, France, Germany, and Slovenia (Crane 1999).

A bee house constructed around 1820 near Richmond, Kentucky, is perhaps the oldest beekeeping structure still standing in the United States (figure 13.1). This was not the only such building in the region, however—Isaac Shelby, Kentucky's first governor, kept a bee house on his property, but it has long since disappeared. The Richmond house is made of brick and stands twelve feet tall, seven feet long, and six feet wide. South-facing, it has in its front wall fifteen square holes where the bees could enter and leave the structure. A back door leads to the interior of the bee house, which was once fitted with five shelves upon which the hives rested. One of the original wooden shelves

FIGURE 13.1 A house apiary constructed in 1820 near Richmond, Kentucky. Photograph by the author.

still exists, bearing evidence of the beeswax that once adhered to its surface. The Richmond bee house was typical of the simpler designs of such structures, and was built more than thirty years before Langstroth discovered the bee space and patented his movable-frame hive.

As beekeepers adopted movable-frame systems, the problems with bee houses became more apparent. One significant issue was expense. The initial cost of building a structure to house an apiary was prohibitive for some beekeepers. Moreover, it seemed an unnecessary step for those who found that wooden box hives worked quite well even when exposed to the elements. Bee house proponents argued that if the beekeeper was careful to select inexpensive materials and make the house large enough to accommodate many hives, the per-hive cost would actually be less than buying box hives. Hives used in bee houses did not need to be complete box hives, and indeed many beekeepers used chaff hives, which did not require wooden covers. Bee houses filled with inexpensive, less durable hives could be very economical.

Another problem of bee houses was their interior darkness, which sometimes forced the beekeeper to stand in the doorway to examine the frames (figure 13.2). Bees dropped off the frames as a result and were crushed under the feet of the beekeeper. Some bees would even crawl up and sting the legs of the beekeeper. To calm the bees, the beekeeper might use a smoker, only to have the bee house fill with smoke. All these problems

FIGURE 13.2 A house apiary showing the beekeeper sitting in the doorway to inspect his bees (Root 1888).

could be remedied by installing a window, which would allow light into the house and also provide ventilation. These windows would attract the bees that fell to the floor, so they needed to be fitted with bee escapes: specially designed screens or openings that allowed bees to leave the bee house interior but not return through the same opening. The bees were forced to return via their own hive's entrance. To further avoid the buildup of smoke in the interior of the bee house, some beekeepers put their smokers inside a box fitted with a pipe leading to the outside. This helped to vent any excess smoke.

Advocates of the bee house pointed out that bees in the dark interior of the house rarely stung. In fact, they argued, aggressive hives could be made more docile by placing them inside a bee house, as bees enclosed in a darker space did not fly about or sting as much as bees in full sunlight. Therefore, they claimed, proper bee house procedures required neither a smoker, a veil, nor gloves.

Those who favored bee houses felt that the apiaries' greatest advantage was convenience for the beekeeper. It was recommended that the structures be constructed of such a size that the beekeeper could store tools, empty hive boxes, and honey-filled supers within (figure 13.3). This was meant to reduce the theft of hives, honey, and equipment. The beekeeper could work the hives in all weather without upsetting the bees. No longer was it necessary to paint the hives, or to be bothered by wet grass and weeds. Bee houses were also effective for overwintering colonies, as they could be heated if the weather became harsh.

In spite of these assurances, not all bee house beekeepers were happy with their systems. One writer pointed out that his north-facing hives did not fare as well as the south-facing ones and suffered extensive winter kill. The most common complaint was the potential fire hazard that bee houses posed. Usually made with inexpensive wood, and filled with wax and chaff, bee houses were highly flammable. If the house stored equipment, the

FIGURE 13.3 An Ohio beekeeper standing in front of his bee house, c. 1900 (Kritsky collection).

beekeeper could lose his or her entire operation to a fire. Although beekeepers could obtain fire insurance, that was yet another expense not incurred by beekeepers with outdoor apiaries.

In the end, the 1891 *Beekeepers' Review* concluded that bee houses could be successfully used if proper bee escapes were incorporated and fire insurance purchased. By that point, however, the days of the bee houses in America were already limited. The 1888 edition of *The ABC of Bee Culture* (Root 1888), included five pages of text and illustrations devoted to bee houses, and the introductory information was far from complimentary. In the 1913 edition, the authors concluded, "As a general thing,

FIGURE 13.4 A garden shed converted into a small bee house for two hives. Photograph by the author.

an outdoor apiary is cheaper and more satisfactory than one in a building (Root and Root 1913)."

Although bee houses declined in popularity throughout the early twentieth century, especially in America, they are still in use in many countries. John Hamer of Blackhorse Apiaries near Woking, England, is waging a campaign to encourage more urban beekeeping. In his demonstration bee-yard, he has a large bee house containing several hives. The entrances to the hives are painted in different colors so that the bees will be able to locate their hive upon their return. John has also converted a small garden shed into a bee house (figure 13.4) that holds two hives in addition to garden tools—ideal for a small English garden. Indeed, bee houses are still popular in parts of Europe, and it is possible to purchase them from beekeeping supply companies such as Swienty, located in Sønderborg near the Dutch and German border.

CHAPTER 14

Bee Calendars

BEEKEEPERS HAVE KNOWN FOR OVER SEVENTEEN hundred years that proper beekeeping requires not only the skills to manage bees, but also an understanding of when to apply those skills. The bee calendar, a checklist of activities to be completed by specific dates over the year, provided such instruction for beginning beekeepers. The first record of a bee calendar dates back to 300 CE, when the Roman author Palladius composed a calendar of husbandry information that also included important beekeeping dates (Dadant 1889).

Nearly all of the early English beekeeping writers discussed when certain tasks needed to be done, but only a few organized them into monthly calendars their readers could follow. One of the first to do so was Thomas Tusser (1586), author of *Five Hundred Points of Good Husbandrie*. His book was organized by months and included information on many aspects of farming, including verses on beekeeping. For September, Tusser wrote:

> Now burn up the bees, that ye mind for to hive,
> at midsummer drive them, and save them alive:
> Place hive in good air, set southly and warm,
> and take in due season, wax, honey, and swarm.

Set hive on a plank (not too low by the ground)
where herb and flowers, may compass it round:
and boards to defend it, from north and northeast,
from showers and rubbish, from vermin and beast.

In December, the beekeeper was taught how to "preserve bees":

Go look to thy bees, if the hive be too light,
set water and honey, with rosemary dight;
which set in a dish, full of sticks in the hive,
from danger of famine, ye save them alive.

In the new year, the beekeeper was to:

Watch bees in May,
for swarming await,
Both now and in June,
mark master bees tune.
Take heed to thy bees, that are ready to swarm,
the loss thereof now, is a crowns worth of harm:
let skillful be ready, and diligence seen,
lest being too careless, thou loses thy bees.

Some early bee books, rather than including full bee calendars, simply recommended that some activities be completed at specific points in the year. Southerne (1593) focused on four dates: Christmas was the best time to construct the hives; March 21 was when one cleaned the hive of dead bees; midsummer was time to drive the bees; and at Bartholomewtide (August 24) one should decide which hives to harvest. Nearly a century later, Rusden (1685) added Michaelmas (September 29) to the important dates for beekeepers, noting that swarms taken in May and June should be considered stocks of bees by Michaelmas.

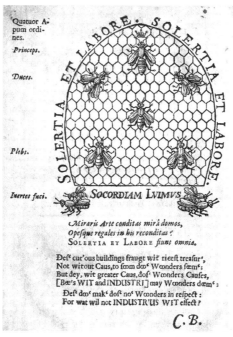

FIGURE 14.1 The frontispiece of Butler's *A Feminine Monarchie*, published in 1634.

The most detailed early bee calendar was also the most unusual. Published by Charles Butler in his 1634 *Feminine Monarchie* (figure 14.1), it began: "The Melissæan year is most fitly measured by the astronomical months, which begin with the Sun's entrance into the several signs of the zodiac, and are therefore called by their names." He included a table (figure 14.2) to help his reader follow the Melissæan year by the astrological signs.

Butler argued that this astrological system enabled the beekeeper to more easily monitor his bees because it included the solstices and equinoxes, taking into account the changing length of the days as the seasons passed. He focused his calendar on the

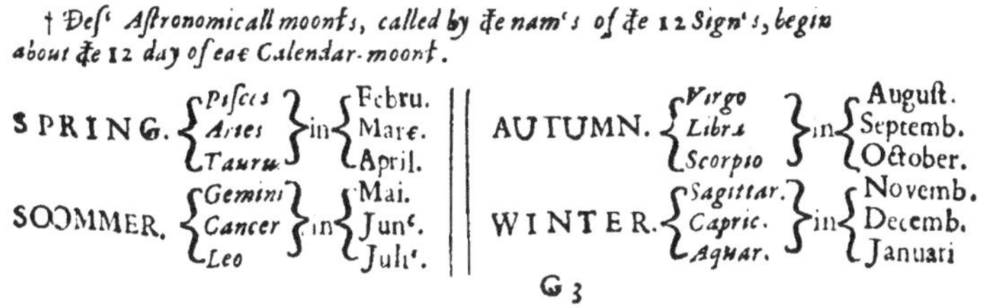

† *Def' Aſtronomicall moonſ, called by ſe nam's of ſe 12 Sign's, begin about ſe 12 day of eaſ Calendar-moonſ.*

SPRING.	*Piſces* *Aries* *Tauru*	in	Febru. Marſ. April.	**AUTUMN.**	*Virgo* *Libra* *Scorpio*	in	Auguſt. Septemb. October.
SOOMMER.	*Gemini* *Cancer* *Leo*	in	Mai. Junſ. Julſ.	**WINTER.**	*Sagittar.* *Capric.* *Aquar.*	in	Novemb. Decemb. Januari

G 3

FIGURE 14.2 The bee calendar from Butler's *A Feminine Monarchie* (Butler 1634).

four seasons as defined by the zodiac, which meant that new seasons began on the 12th day of the month. Thus, spring began on February 12 and ended on May 11. Such a system was broad enough to incorporate seasonal variations, such as an early or a late spring. Of the seasons, Butler wrote:

> But the four quarters of the bee's year begin one month sooner than the astronomers. For the Spring or first quarter begins with Pisces, when the sun begins by his quickening heat to revive the flowers, which all the dead of winter lay buried in the ground; and the bees having tasted thereof, begin to breed, and to increase their companies, for the fruits of ensuing summer: which from the former summer hitherto have daily decreased: the other spring months are Aries and Taurus.
>
> The summer likewise continues with Gemini, Cancer, and Leo, most rich and plentiful in flowers and dews: with the multiplied bees do nourish their cells against the penury of winter.
>
> The autumn of harvest, has Virgo, Libra, and Scorpio: in which the bee masters and the master bees do reap the fruits of many bee's labors.
>
> And the winter consists of three still months in which the bees live altogether upon their summer store and get nothing.

As with previously mentioned calendars, Butler's instructions were written with skep beekeeping in mind and therefore included some activities specific to skep hives. For example, during the time of Leo, the beekeeper is instructed to "kill the drones of those stalls [hives] you mean to take, with a drone-pot cloomed [attached by mud or dung] to the door." This was done by placing a wire grate in the entrance of the hive, and reducing its opening to prevent the drones from getting back into the hive. This action lowered the population of the drone bees, which was (incorrectly) assumed to reduce the bees' consumption of honey and thus increase the yield. The reduced number of drones was also incorrectly believed to delay the next year's swarming.

Butler's zodiac-based bee calendar did not catch on with other authors. In 1681, William Mather returned to the traditional method of listing activities by month in his book, *A Very Useful Manual or the Young Man's Companion* (figure 14.3).

Moses Rusden (1685), an early proponent of the wooden box hive, used a bee calendar as a guide to aid his readers in their transition from straw skeps to octagonal box hives. His bee calendar focused on the management of straw skeps, but also included information on when and how to move bees to octagonal boxes. This system had the advantage of being familiar to the experienced skep beekeeper while aiding in the transition to wooden octagonal box hives.

As the nineteenth century witnessed many changes in the design of the hive, the bee calendar continued to serve as a guide to beekeeping practice. Langstroth (1859) felt that the bee calendar was critical for the novice beekeeper and provided detailed monthly instructions for management of his movable frames. For example, he wrote of August: "If any colonies are so full of honey, that they have not room for raising brood, some combs should now be removed. If the caps of the cells are carefully sliced off with a very sharp knife, and the combs laid over

(279)

FIGURE 14.3 William Mather's 1681 bee calendar.

Memorandums.

1. LAy falt on the ftool, and afhes on the Crown once a year.

2. Drefs the ftools in *February*.

3. Feed them but in *April* with wet Sugar, if rainy weather come.

4. The Queen Bee is often found on the ground under a Hive of two fwarms, and after the laft fwarm.

5. Bees age fcarce two years.

6. Help them to kill the Drones in *July*, or at other times, to prevent fwarming,

7. *June* the beft Month for Bees.

8. They breed moft in wet weather, and almoft all Summer.

9. When they caft out young white Bees, the flock is good.

10. Take no Hony before *Auguft*.

11. Remove Bees in *January*, and to carry them many Miles, wrap a Cloath about them, and turn them upfide down on Straw in a Cart.

12. To try to make them fwarm, lay ftinking weeds under them that they lie not out, but above the door, and the Sun fhine not on them but at the door; if this do not make them fwarm, rare them two Inches, and they often fwarm in two or three days after; do this about the middle of *June*. 13. In

a vessel, in some moderately warm place, and turned once, most of the honey will drain out of them, and they may be returned to the bees, to be filled again."

One of the most complete bee calendars was found in Cheshire's (1888) second volume of *Bees and Beekeeping*. England's beekeeping industry was still transitioning from skeps to movable-frame hives, and Cheshire included information for both kinds of hives, such as when it was best to purchase skeps for transferring into boxes (September), when stocks should be returned from the moors (October), when to sow certain plants (March–May), and when to take heather honey (August).

Bee calendars, unlike beehives, are not made of straw or wood. Based on folk philosophies, they also changed during the transition from skeps to frames. Butler's zodiac-based system and Tusser's poetry portrayed beekeeping as an art, but the innovation of a bee calendar by which beekeepers managed hives reflects beekeeping's slow evolution into a science.

CHAPTER 15

Beekeeper's Paraphernalia

To someone unfamiliar with apiculture, a beekeeper decked out in a white bee-suit and veil with smoker in hand might appear to be working with hazardous waste. Beekeepers now take protective clothing and smokers for granted, but they were not always the norm. It seems the ancient Egyptians worked their tube hives without any protective gear at all, using an incense burner as a smoker. The first-century Roman writer Columella described a pottery smoker with a large opening at one end and a narrow opening at the other. The beekeeper would blow air into the large opening, pushing smoke out of the narrow side (Crane 1999). By the fourth century, Nonnus wrote that beekeepers would cover themselves from head to toe (probably with their tunics) to protect against beestings (Crane 1999).

BEE-SUITS

Bee-suits likely originated in fifteenth-century Europe (figure 15.1) (Crane 1999). These early suits covered the head and shoulders with the primary goal of protecting the face. They restricted

FIGURE 15.1 A bee-suit
from a 1532 wood cut
(Kritsky collection).

visibility and dexterity, however, and beekeepers considered them a last resort to protect themselves against stings. A more primitive sting repellent was attention to one's personal hygiene. In his *Feminine Monarchie* (1704), Charles Butler wrote that once a bee had stung through the clothing, other bees would smell the "scent" of the sting and be enticed to do the same. Given the bees' sensitivity to scent, Butler suggested that one could quiet bees by paying attention to one's odor. He wrote, "There are several things that will vex and anger the bees, as your having any ill offensive scent about you when you approach them; . . . all nastiness in apparel is offensive to them, as also an unsavory breath; natural, or by reason of eating any strong things, as leeks, garlic, onions, rue, etc." In short, by staying clean, a beekeeper reduced the likelihood of being stung. Butler added, however, that if the beekeeper had to work angry bees, he should "put a thin cloth or hood over [his] face," as it was the first area the bees would likely target.

Like Butler, James Bonner (1789) was concerned about presenting bees with a pleasant smell. He wrote, "[C]ome not among them in a rash hasty manner; neither must you come puffing and blowing, nor with bad smells about you." Bonner also claimed that bees were offended by head, beard, or eyebrow hair.

These theories did not stop Bonner from making use of protective gear, however. A Scot, he called his bee-dress a "harness," referring to a suit of armor, and in fact used the two terms interchangeably. The harness consisted of a square yard of canvas, "wove very slack," that permitted him to see and breathe. He placed this shroud over his head, drew it together around his neck, and held it in place with a garter. In addition, he wore gloves and tied his coat sleeves above them to stop bees from crawling inside his coat.

By the end of the eighteenth century, bee books contained discussions of "bee-dress," or clothing specifically made to protect against bees. John Keys (1796) included detailed instructions for the preparation of protective gear using bolting cloth. Bolting cloth was used to sift flour, and its open weave allowed beekeepers to see through it, improving on earlier hoods that limited vision. Keys instructed his readers to sew the cloth to an old hat with a brim that had been cut back to two and a half inches in width. The cloth would extend down for at least a foot on all sides of the hat, the bottom edge finished with a tape or string to tie the material around the neck and under the chin. To further protect the face and neck from beestings, Keys recommended that oiled linen be stitched to the front. Oiled linen was prepared by soaking linen in linseed oil and wringing out the excess. After it dried, the process was repeated to give the fabric more strength and provide a greater barrier against bees. If this

FIGURE 15.2 The bee-dress
recommended by T. B. Minor
(1849).

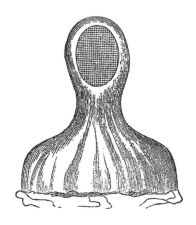

wasn't enough, a hood could be made of the same fabric and put over the hat when necessary; the brim would keep it away from the face. Keys also treated leather gloves with oil, which, once dried, made them waterproof.

Another material commonly used in making protective gear was wire mesh. Starting in the late eighteenth century, wire mesh was used as a means of protecting the face from stings. Woven metal had the benefit of being very stiff, which guaranteed that the distance between the wire and the face would remain constant. Untreated fabric could drape and touch the face of the beekeeper, providing opportunities for bees to sting the face. T. B. Minor (1849) recommended a muslin cowl with a wire mesh insert as the best bee-dress (figure 15.2).

W. C. Harbison, an American beekeeper (1860), noting that an "ounce of prevention is better than a pound of cure," pointed out that many beekeepers failed to properly manage their hives because they were fearful of getting stung. This was especially true in his home state of California, because temperatures could get quite hot, making the bees more aggressive and likely to sting. Given the temperature, Harbison preferred a "veil" made of silk bobbinet attached to a cheap summer hat (figure 15.3). Though he tried hats made of wire cloth, he found they did not

FIGURE 15.3 A bee-veil that could be attached to a hat (Harbison 1860).

filter out the sun's rays, giving him sunburn. As the hats absorbed energy from sunlight, they also became very hot.

By the second half of the nineteenth century, the use of a veil and hat was common practice in both England and the United States. Alfred Neighbour (1878) sold two types of "protectors" to be used with hats. One of his innovations in bee dress was the choice of black netting, which reflected less light and provided improved visibility. The netting was cut in the shape of a large bag that fit over a gentleman's or lady's hat. Furnished with two elastic-edged armholes, the veil was held together at the waist with a large elastic band (figure 15.4). Neighbour also sold a cheaper model with an elastic band that secured the veil to a wide-brimmed hat; its bottom edge was tucked into the beekeeper's jacket. This kind of design is still used—indeed, during the late nineteenth century, Neighbour and other purveyors of beekeeping equipment were selling veils that would not look out of place if worn today.

Beekeepers' hats were more variable, reflecting the preferences of the individual beekeepers. It was possible to find beekeepers using their veils with caps, flat straw boaters, rounded bowler hats, old fedoras, large-brimmed straw hats, and even, remarkably, top hats (figure 15.5). Improved fabrics and construction are used today to make bee suits and bee jackets with built-in veils and hats, but

FIGURE 15.4 The larger bee-protector sold
by Neighbour (1878).

FIGURE 15.5 An English beekeeper in his top hat and veil (from a
magic lantern slide, Kritsky collection).

FIGURE 15.6 The author in his bee-suit. Photograph by J. Griffith.

with design echoes of the past—even modern suits share similarities with designs of the sixteenth century (figure 15.6).

THE BEE SMOKER

Humans have known about smoke's ability to quiet bees for centuries, if not millennia; Mesolithic rock paintings suggest the use of smoke by ancient bee hunters. Why smoke quiets bees has been the subject of much speculation. Langstroth (1859), upon observation that bees filled their crops with honey in response to smoking, concluded that smoke must frighten bees, believing frightened bees to fill crops with honey. Cheshire (1888) rejected part of Langstroth's idea. He pointed out that he had been stung many times by bees from a hanging swarm, showing that the bees' being filled with honey had nothing to do with their submissive behavior. Rather, he argued that bees exposed to smoke did not sting because they were "terror-struck." Recent experiments have proven Cheshire wrong, showing that smoke fouls the bees' sensory receptors and causes some to engorge with honey (Shimanuki et al. 2007). The evolutionary reason is likely tied to honey bees' survival strategy—their inclination for inhabiting hollow trees required them to respond to the scent of smoke from a forest fire.

Over time, beekeepers developed specific practices and devices to control and more safely work with the bees. In the early days of skep beekeeping, it was common to kill the bees in a brimstone pit. Butler (1634) gave instructions for digging a pit that was the same diameter as the skep, lighting a sulfur stick placed in the bottom of the pit, and putting the hive over it. Within a few minutes, the bees would succumb to the fumes and fall into the pit. Because it might not be easy to dig a pit near all of the skeps, Butler gave the option of using a movable "pit" made of an elm trunk ten inches tall and 18 inches in diameter. The top and the bottom of the trunk would be dug out, in part to lighten the weight but also to provide a receptacle for the dying bees. The brimstone stick was lit and placed in the upper concavity, with the skep placed on top and draped with cloth to seal in the smoke.

The fuel used in these pits was sulfur, but there were less toxic options as well. Dried puffballs, mentioned by Butler (1634), became very popular for subduing bees as beekeepers stopped killing them at harvest time. The puffball fungus, *Lycoperdon giganteum*, grew in pastures and was harvested in the autumn (Taylor 1860). They were gathered when almost ripe and sun- or oven-dried to preserve them. Though the puffball was preferred, if they were unavailable other fungi could be substituted.

The puffball was also very effective for uniting hives. Lighting an egg-sized piece of puffball, the beekeeper would place it in a box. The skep was then placed over this box and draped with cloth. At first, the hive would sound as if it were in an uproar, but the sound would quickly subside, and quieted bees would drop out of the hive into the box. It was common to hit the sides of the skep to dislodge any bees clinging to the comb. The bees were then dumped onto a sheet and sorted with a large feather in order to locate the queen for removal. The subdued bees, placed into a bowl, were sprinkled with sugared ale, honey water, or peppermint to ease their integration into the new hive. The skep to be joined with the quieted one was placed over the bowl, and its workers would start to "lick" the subdued bees. This provided the quieted bees with the scent of their new hive and aided in their acceptance. If the original hive was weak, this procedure could be repeated to unite it with another hive (Taylor 1860 and Neighbour 1866).

Puffball use resulted in the development of dedicated equipment. Thomas Nutt invented an elaborate puffball "smoker," which looked like a pepper shaker, into which the lit puffball was placed. This was attached to a rod above a funnel-shaped bottom (figure 15.7), which was combined with a sheet that hung below the skep to catch the drugged bees and funnel them into a container (figure 15.8) (Taylor 1860).

Tobacco was another common smoker fuel in the days of skep beekeeping. Beekeepers found that blowing a few puffs of smoke into their hives would suffice to calm the bees (figure 15.9). Some veils, still in use in parts of Europe today, were even designed with an opening for the stem of the small "pipe style" smoker (figure 15.10). This worked well with small skeps, but movable-frame hives needed much more smoke than pipes or other small smoke devices could produce. For larger hives one needed a bellows.

The use of a bellows to aid in smoke production began with skep beekeeping. These two-to three-foot-long smokers consisted of a large cylinder with a tube running out of either end (one for

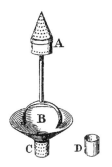

FIGURE 15.7 Thomas Nutt's puffball fumigator. The dried puffball, called a puk, was placed in the top chamber (Taylor 1860).

FIGURE 15.8 Nutt's fumigator fitted for use on a skep (Taylor 1860).

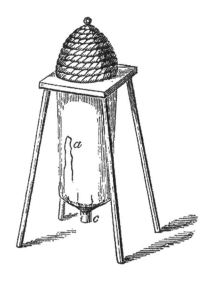

FIGURE 15.9 Cigars were often used as a means of quieting bees (Bessler 1921).

delivery of the smoke and the other for a bellows) (figure 15.11) and required both hands to use. Their drawback was that they ran out of smoke relatively quickly.

The precursor to the modern bellows smoker was invented by Moses Quinby in 1873 (figure 15.12a–d). This early smoker

FIGURE 15.10 A small pipe smoker in the IBRA's collection in the International Beekeeping Museum at Eeklo in Belgium. Photograph by the author.

FIGURE 15.11 An early bellows smoker being used with a skep (Cotton 1842).

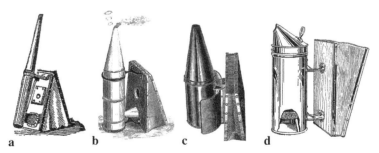

FIGURE 15.12 The evolution of the modern smoker: Quinby's first smoker (a), Quinby's improved smoker (b), Bingham's smoker (c), and Woodman's smoker (d) (Root 1888 and Pellett 1938).

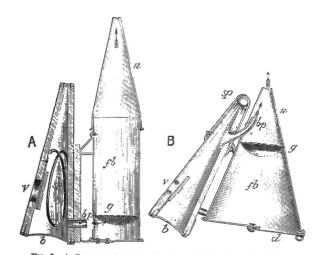

FIG. 3.—A, BINGHAM SMOKER. B, CLARK'S COLD DRAUGHT SMOKER
(Sectional View, one-fifth actual size).
b, Bellows; *v*, Valve; *sp*, Spring; *fb*, Fire-box; *d*, Door to Fire-box; *g*, Grating;
n, Nozzle; *bp*, Blast-pipe.

FIGURE 15.13 Comparing the internal construction of a Bingham smoker with the Clark's cold blast smoker (Cheshire 1888).

had the bellows alongside the fuel chamber, which dramatically reduced its size; however, users found that it ran out of smoke right when it was needed most, when the hive was opened. T. F. Bingham's design in 1877 included a space below the fuel chamber for the air to enter (figure 15.13a), giving the beekeeper more control of the fuel. Although it kept fuel burning for a longer period of time, Bingham's modification proved controversial. The design resulted in air blowing through the fuel, and this hot air blew directly onto the bees. To correct this flaw, Norman Clark developed the cold draught smoker (figures 15.13b and 15.14), which directed the blast of air above the fuel chamber, pushing smoke out of the smoker rather than through the fuel. Unfortunately, the cold draught smoker did not provide the fuel control of the Bingham smoker (Pellett 1938, Root 1888, and Cheshire 1888), and it eventually fell out of favor.

FIGURE 15.14 A Clark's cold blast smoker. Photograph by the author.

The last step in the smoker design was made by Bingham's successor, A. G. Woodman. By reversing the orientation of the bellows, he made it easier for beekeepers to handle, no longer requiring the user to reach under the smoker to get to the wider end of the bellows (figure 15.12d). This popular design has seen only a few improvements since its modification in early twentieth century (Pellett 1938).

Changes in smoker design have always followed changes in hive design. Skeps needed only limited smoke for inspection, and pipes were sufficient. The larger movable-frame hives demanded a longer-lasting source of smoke, and the bellows smokers appeared within 25 years. Today, the smoker is a common symbol of the beekeeper. It is, as Cheshire wrote in 1888, "the beekeeper's talisman," giving its possessor seemingly magical powers over his bees.

CHAPTER 16

The End of Innovation

An EXAMINATION OF BEEKEEPING BOOKS FROM the eighteenth and nineteenth centuries reveals the creative spirit of the era's beekeepers. By contrast, the twentieth century witnessed a steep decline in new beehive design. Only a handful of beehive models are currently used in England, and the vast majority of hives in America are of the same design, varying mostly in number of frames and frame sizes. What stifled the prior creativity? Why the sudden loss of innovation? The answer: simple economics. The movable-frame hive dramatically increased honey yields, but the expense of the hives—combined with the steep prices for the new extractors—made it extremely unappealing for beekeepers to change equipment.

In the early nineteenth century, harvesting honey was a slow process that had changed little since Roman times: one simply waited for the honey to drip out of the combs. Honey drainers from the 1840s (figure 16.1) were made from two oval tin vessels measuring 18" long, 7" wide and 5" deep. The top vessel was perforated on the bottom with holes about 1/16" in diameter, and it nested inside the bottom vessel, resting on a beaded rim. The bottom vessel was equipped with a spigot to drain the honey. A piece of muslin placed between the two vessels acted as a sieve (Dunbar 1840).

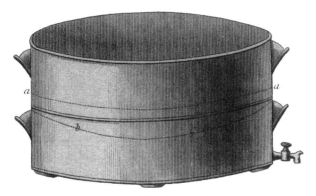

FIGURE 16.1 A honey drainer (Dunbar 1840).

Honey was harvested as soon as the bees had been driven from the skep and the comb cut away. Extraction took place in a heated room, as the warm air would help the honey flow more quickly. Sealed comb was cut through the plane of the comb and placed cut side down so that the honey could drip out. The honey flowed through the holes in the top vessel and collected on the muslin sieve. The material trapped any small pieces of wax and other matter, allowing only pure honey through. Once filtered, it could be dispensed into a jar through the spigot (Dunbar 1840).

Major Franz Ealer von Hruschka, an Austrian living in Italy, discovered the centrifugal process for extraction in 1865 after watching his son play with a small piece of honeycomb he had given him. The child put the honeycomb in a basket and swung it around his head. When Hruschka looked inside the basket, he discovered that the honey had been flung out of the comb onto the sides of the basket—or so the story goes (Cheshire 1888). Using this principle, Hruschka invented a metal extractor, a square funnel with a wire handle that held a single frame of honey (figure 16.2). The handle was placed on a wooden rod held by two people, one at each end. Once attached, the extractor was spun, much like a jump rope. As the extractor was spun, the honey collected inside the extractor and was drained out from

FIGURE 16.2 Hruschka's
first extractor (Pellet 1938).

FIGURE 16.2 Hruschka's first extractor (Pellet 1938).

the tube at the bottom. This simple extractor was described in 1865. A year later, Hruschka enlarged the extractor to hold eight frames.

Once the centrifugal process of extraction was discovered, others started experimenting. By 1890, extractors that could hold up to eight frames placed tangentially along the side of the extractor were common (figures 16.3 and 16.4). The invention of the radial extractor, in which the frames were arranged with the top bars radiating outward from the center (figure 16.5), resulted in extractors that could handle as many as 120 frames at a time. The efficiency of these extractors greatly expanded the honey industry.

It was not a coincidence that the invention of efficient extractors occurred as the Langstroth hive was gaining in popularity. Just as the smoker changed to accommodate the new style of beekeeping, the extraction of honey by mechanical means was certain to follow. However, an extractor was an expensive investment. The marketplace demanded cheaper models, which required the mass production of extractors. Working against this was the large variation in frame size. A mass-produced extractor would not be able to accommodate all the frames in use. It was in the interest of lower prices, then, that frame sizes were standardized. In England, a committee was formed to evaluate

FIGURE 16.3 A tangential extractor (Pellet 1938).

FIGURE 16.4 Two beekeepers preparing frames for extraction using a tangential extractor, c. 1900 (from a lantern slide in the Kritsky collection).

FIGURE 16.5 A radial extractor (Pellet 1938).

and standardize frame sizes. In America, purveyors of beehives sold extractors to match the size of the frames they were selling. There were eight popular frame sizes in 1890; however, the A. I. Root Co. sold extractors that fit only six of the sizes. Within 40 years, the number of frame sizes was down to four.

By the early twentieth century, beekeepers had invested heavily in equipment, and purveyors of beekeeping supplies sold frames, bottom boards, and covers that were interchangeable, so that equipment could be maintained more cheaply. By the 1920s, standardization had locked beekeepers into a design and size. This one-size-fits-all approach was detrimental to innovation in hive design. A beekeeper who wanted to adopt a newly invented hive would essentially have to start from scratch. New beehives would use new frame sizes, which would in turn require a new extractor.

This decline in beekeeping innovation is apparent when you compare beekeeping supply catalogs published in the last 100 years.

The contents of the 1890 Root catalog seem diverse and exotic, including several different types of hives and several frame sizes, the descriptions indicating which extractors would work with which type of frame. The 1929 Lewis catalog, by contrast, looks like it might have been printed last week. The hives offered are all still available and the choice of frames is limited to three. A 2008 Dadant catalog has a few novel items for sale—hives made of Styrofoam and stainless steel extractors—but their design is essentially the same as those sold in the 1920s. The one notable difference between the older and newer catalogs is that the latter include a wide array of antibiotics and chemicals for treatment against disease and mites.

That's not to say that beekeepers of days gone by did not worry about disease and pests. Skep beekeepers kept a watch out for foul-smelling colonies, and the Isle of Wight disease, believed to be caused by a mite, devastated beekeeping in the early twentieth century. Even the current beekeeping concern, Colony Collapse Disorder (CCD), may have been observed more than two centuries ago. Bevan (1827) described CCD-like symptoms when he documented a major die-off in England. He wrote, "In the winter of 1782–3, a general mortality took place among the bees in this country, which was attributed to various causes: want of honey was not one of them; for in some hives considerable stores were found, after the bees were gone."

The evolution of the modern beehive illustrates how beekeepers modified and sometimes improved the beehive during the 1,500 years since straw skeps were introduced (figure 16.6). The invention of supering helped to control swarms and increase yields. The introduction of bars facilitated the beekeeper's control of the combs. The transition to wooden hives, though it met firm resistance initially, took hold when combined with wooden frames in the true movable-frame hive.

Innovation can address apicultures' concerns as illustrated with the relatively recent invention of the top-bar hive (figure 16.7d).

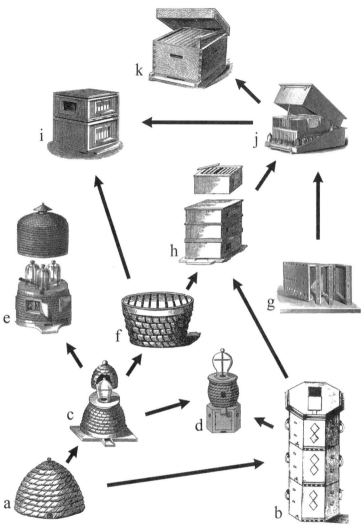

FIGURE 16.6 The innovations in the evolution of the modern beehive: (a) a straw skep (Bagster 1838), (b) a wooden octagonal box hive (Gedde 1675), (c) Milton's Cottage Hive (Milton 1823), (d) Thorley's Depriving System (Bagster 1838), (e) Neighbour's Improved Cottage Hive (Neighbour 1878), (f) Golding's Grecian bar hive (Filleul 1856), (g) Huber's Leaf Hive (Cheshire 1888), (h) Bevan's bar hive (Taylor 1860), (i) Neighbour's new straw frame hive (Neighbour 1878), (j) Langstroth's original movable frame hive (Langstroth 1859), (k) modern Langstroth hive (Root and Root 1923).

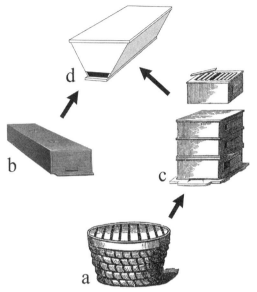

FIGURE 16.7 The innovations in the evolution of the top-bar hive:
(a) Golding's Grecian bar hive (Filleul 1856), (b) Janscha's horizontal
hive (Janscha et al. 1900), (c) Bevan's bar hive (Taylor 1860),
(d) top-bar hive (drawing by the author).

These hives, developed by C. J. Tredwell and Peter Paterson
in 1965, are horizontal hives (figure 16.7b) that combine the prin-
ciples of the Grecian hive (figure 16.7a) with the Bevan bar hive
(figure 16.7c). The wooden top-bar hive is constructed with sides
that slope down from the top to a narrow bottom. About 20
wooden bars, below which the bees build their combs, are placed
across the top. The key to this hive is the 120° angle formed
between the bottom and the sides. Apparently the angle is wide
enough to trick the bees into treating the sides as a hive bottom,
and thus they do not attach any comb to the sides or bottom,
resulting in a bar with an easily removable comb. The hive can
be built from just a few boards, and the honey can be harvested
without the use of an expensive extractor, making them ideal for
beekeepers in developing countries with hot climates (Shimanuki

et al. 2007). These hives address both the climatic and economic issues facing new beekeepers in east Africa.

Innovative beehives can incorporate the best of several old hives. The new hive sold by the British firm Omlet illustrates this combination. Their hive, called the Beehaus, is a horizontal hive made of triple-insulated plastic that is easy to clean, thus reducing disease, and easy to inspect for parasites. It has a mesh floor to provide ventilation, and its length promotes brood production and can be divided to help control swarming. Small, easy-to-handle honey supers sit above the queen excluders, promoting the production of quality honey. These needs have been incorporated into hives of the past. Chaff hives were also insulated, straw skeps could be easily replaced with new skeps to help lessen disease, and the straw provided good ventilation. A long horizontal hive to control brood production and swarming was the purpose of Janscha's horizontal hives, and small glass jars or skeps above the main skep helped the small beekeeper to produce quality honey. Finally, the Beehaus was aesthetically designed to promote urban beekeeping and to be a pleasing addition to the urban garden, which was the desire of the ornate hives of the nineteenth century. The Beehaus is too expensive to answer the needs of commercial beekeepers, but it shows how old technology can lead to innovative hives.

Beekeeping has been practiced for nearly 5,000 years. Around the world today, you can find beekeepers using horizontal hives, logs, simple boxes, skeps, top-bar hives, and hives with movable frames. Each one of these hives has survived because it has been adapted for the specific situation in which it is used. But these hives also have problems that limit honey production, foster diseases, or incur excessive costs.

Scientists have recently worked out the honey bee genome, and understanding the genetics of honey bees may be the greatest innovation in apiculture since Langstroth's discovery of the bee space. We do not know where molecular biology will take beekeeping, but one thing is certain. In spite of high-tech biology, we are keeping our bees in "old" hives. Are we really using "perfect" hives? Because we have stopped inventing hives, we really do not know. We have nearly forgotten about the old hives and ways of working them. If beekeeping is to survive the challenges of the molecular age, we need to consider the advantages and creative solutions presented by strange old beehives and, as exemplified by the recent innovation of the top-bar hive, use that knowledge to start inventing again. If we do, we may find the next perfect hive.

Bibliography

Abbott, C. N. 1876. The standard frame. British Bee Journal 3: 221–222.

Anonymous. 1851. The great gathering of the industrious bees. 1851. Illustrated London News September 20, 1851, No. 518, Vol. 19: 359–360.

Anonymous. 1851a. The official and descriptive catalog. Vol. 1. Spicer Brothers, London.

Anonymous. 1876. The Centennial honey show. American Bee Journal 12: 297–298.

Argus. 1878. The Paris Exhibition. American Bee Journal 14: 299–302.

Bagster, S. 1838. The management of bees. Saunders and Outley, London.

Bessler, J. G. 1921. J. G. Bessler's Illustriertes Lehrbuch der Bienenzucht. W. Kohlhammer, Berlin.

Bevan, Edward. 1827. The honey bee, its natural history, physiology, and management. Van Voorst, London.

Bevan, E. 1838. The honey-bee. Van Voorst, London.

Bonner, J. 1789. The bee-master's companion, and assistant. J. Taylor, Berwick.

Brown, R. 1994. Great masters of beekeeping. Bee Books New & Old, Somerset, England.

Burr, S. J. and S. D. V. Burr. 1877. Memorial of the International Exhibition. S. Stebbins, Hartford, CT.

Butler, C. 1634. The feminine monarchi or the histori of bees. William Turner, Oxford.

Butler, C. 1704. The feminine monarchie, or the history of bees. A. Baldwin, London.

Butterworth, B. 1892. The growth of industrial art. Government Printing Office, Washington, D.C.

Cheshire, F. R. 1886. Bees & beekeeping; scientific and practical. L. Upcott Gill, London. Vol. 1.

Cheshire, F. R. 1888. Bees & beekeeping; scientific and practical. L. Upcott Gill, London. Vol. 2.

Cook, A. J. 1884. The bee-keepers' guide or manual of the apiary. A.J. Cook, Lansing, MI.

Cotton, W. C. 1842. My bee book. J.G.F. & J. Rivington, London.

Cowan, T. W. and W. B. Carr. 1891. The late Mr. Alfred Neighbour. British Bee Journal 19: 1–2.

Cowan, T. W. 1881. British bee-keepers guide book. Houlston & Sons, London.

Cowan, T. W. 1913. British bee-keepers guide book. 21st ed. Madgwick, Houlston & Co., London.

Crane, E. 1983. The archaeology of beekeeping. Cornell Univ. Press, Ithaca, NY.

Crane, E. 1999. The world history of beekeeping and honey hunting. Routledge, New York.

Cumming, J. 1864. Bee-keeping. By "The Times" bee-master. Sampson Low, Son & Marston, London.

Dadant, C. 1889. Langstroth on the hive and the honey bee. Chas. Dadant & Son, Hamilton, IL.

Doddridge, J. A. 1813. A treatise on the culture of bees. St. Clairsville, OH.

Dunbar, W. 1840. The natural history of bees. Lizars, Edinburgh.

Duruz, R. M. and E. E. Crane. 1953. English bee boles. Bee World 34(11): 209–224.

Dzierzon, J. 1882. Dzierzon's rational bee-keeping. Houlston and Sons, London.

Fenwick, H. 1976. Scotland's castles. Robert Hale, London.

Filleul, P. V. M. 1856. The cottage beekeeper. C. M. Saxton & Co., New York.

Forster, E. S. and E. H. Heffner. 1969. Lucius Junius Moderatus Columella on agriculture. Vol. 2. W. Heinemann, Ltd., London.

Galton, D. 1971. Survey of a thousand years of beekeeping in Russia. Bee Research Association, London.

Gedde, John. 1675. A new discovery of an excellent method of bee-houses and colonies. D. Newman, London.

Gee, H. 1984. Glossary of building terms from the conquest to c. 1550. Frome Historical Groupe, Frome, England.

Harbison, W. C. 1860. Bees and bee-keeping: a plain, practical work. C. M. Saxton, Barker, & Cop., New York.

Hartlib, Samuel. 1655. The reformed commonwealth of bees. Giles Calvert, London.

Heddon, J. 1885. Success in bee-culture. Times Print, Dowagiac, MI.

Herrod-Hempsall, W. 1930. Bee-keeping new and old described with pen and camera. Vol. 1. British Bee Journal, London.

Herrod-Hempsall, W. 1937. Bee-keeping new and old described with pen and camera, Vol. 2. British Bee Journal, London.

Hibberd, S. 1856. Rustic Adornments for homes of taste, and recreations for townfolk in the study and imitation of nature. Driffield, London.

Hibberd, S. 1987. Rustic Adornments for homes of taste. Reprint of 1856 ed. National Trust, London.

Hicks, T. C., R. S. Fouts, and D. H. Fouts. 2005. Chimpanzee (*Pan troglodytes troglodytes*) tool use in the Ngotto Forest, Central African Republic. American Journal of Primatology 65 (3): 221–237.

Hill, Thomas. 1568. The profitable arte of gardening. Thomas Marsne, London.

Hill, Thomas. 1577. The gardeners labyrinth. Henry Bynneman, London.

Howison, J. 1819. An essay on the management of bees, with an account of some curious facts in their history. Memoirs Caledonian Horticultural Soc. 2:121–133.

Huber, F. 1808. New observations on the natural history of bees. Anderson, Longman, Hurst, Rees, and Orme, Edinburgh.

Hunter, J. 1875. Manual of Bee-keeping. Hardwicke and Bogue, London.

Hutchinson, W. Z. 1891. The beekeepers' review. 4: 132, 186, 201, 203, 205–207, 211, 214–215, 233, 234, 237–241, 245, 293.

IBRA (International Bee Research Association) 1979. British bee books: a bibliography 1500–1976. IBRA, London.

James, E. L. B. 1950. Bee-keeping for beginners and others. E.L.B. James, Birmingham.

Janscha, A., J. Münzburg, H. Nufer, and F. J. Untergasser. 1900. Die Bienenzucht von Janscha. Selbstverlag der Herausgeber, Deffingen and Holbruck.

Jones, G. F. and R. Wilson. 1981. Detailed reports on the Salzburger emigrants who settled in America, edited by Samuel Urlsperger. Vol. 6. University of Georgia Press, Athens, GA.

Keys, J. 1780. The practical bee-master. J. Keys, London.

Keys, J. 1796. The antient bee-master's farewell. G. G. and J. Robinson, London.

Kritsky, G. 1991. Lessons from history: the spread of the honey bee in North America. American Bee Journal 131:367–370.

Kritsky, G. 2007. The Pharaohs' apiaries. Kmt: The modern journal of ancient Egypt 18(1): 63–69.

Kritsky, G. and R. Cherry. 2000. Insect Mythology. Writers' Club Press, Lincoln, NE.

Langstroth, L. L. 1852. Beehive. United States Patent Office. Letters Patent No. 9,300.

Langstroth, L. L. 1853. Langstroth on the hive and the honey-bee; a beekeeper's manual. Hopkins, Bridgman and Co., Northampton, MA.

Langstroth, L. L. 1859. A practical treatise on the hive and the honey-bee. A. P. Moore and Co., New York.

Lawson, W. 1618. A new orchard and garden with the country housewife's garden with hearbes of common use as also the husbandry of bees, with their several uses and annoyances. B. Alsop for R. Jackson, London.

Leggett, M. D. 1874. Subject matter index of patents 1790–1873, inclusive. Government Printing. Washington, D.C.

Levett, John. 1634. The ordering of bees: or, the true history of managing them from time to time, with their hony and waxe, shewing their nature and breed. Thomas Harper for John Harrison, London.

Loudon, J. C. 1835. Encyclopaedia of gardening. Longman House, London.

MacGibbon, D. and T. Ross. 1887. The castellated and domestic architecture of Scotland, Vol. 2. David Douglas, Edinburgh.

Mather, W. 1681. A very useful manual or the young man's companion. T. Snowden, London.

Mazar, A. and N. Panitz-Cohen. 2008. To what god? Biblical Archaeology Review 34(4): 40–47, 76.

Milton, J. 1823. On bees. J. Milton, London.

Minor, T. 1804. The experienced bee-keeper. T. Collier, Woodbury, Litchfield, CT.

Minor, T. B. 1849. The American beekeeper's manual. C.M. Saxton & Co., New York.

Minor, T. B. 1857. The American beekeeper's manual. C.M. Saxton & Co., New York.

Muth, C. F. 1881. Practical hints to bee-keepers. Muth, Cincinnati, OH.

Naile, F. 1976. America's master of bee culture: the life of L.L. Langstroth. Cornell Univ. Press, Ithaca, NY.

Neighbour, A. 1866. The apiary. 2nd ed. Kent and Co., London.

Neighbour, A. 1878. The apiary; bees, beehives, and bee culture. George Neighbour and Sons, Kent and Co., London.

Nutt, Thomas. 1832. Humanity to honeybees. H. and J. Leach for the author, Wisbech.

Pellett, F. C. 1938. History of American beekeeping. Collegiate Press, Inc., Ames, IA.

Pettigrew, A. 1870. The handy book of bees. William Blackwood and Sons, Edinburgh and London.

Potter, B. 1908. The tale of Jemima Puddle-Duck. R. Warne & Co. Ltd., London.

Quinby, M. 1867. Mysteries of beekeeping explained. Orange Judd & Co., New York.

Ransome, H. 1986. The sacred bee in ancient times and folklore. Bee Books New & Old, Bridgwater, England.

Richardson, H. D. and J. O. Westwood. 1852. The hive and the honey-bee. Wm. S. Orr & Co., London.

Root, A. I, 1877. The ABC of bee culture. A. I. Root, Medina, OH.

Root, A. I, 1883. The ABC of bee culture. A. I. Root, Medina, OH.

Root, A. I, 1887. The ABC of bee culture. A. I. Root, Medina, OH.

Root, A. I, 1888. The ABC of bee culture. A. I. Root, Medina, OH.

Root, A. I. 1890. Bees and honey illustrated catalog and price list. A. I. Root, Medina, OH.

Root, A. I. and E. R. Root. 1913. The ABC and XYZ of bee culture. A. I. Root, Medina, OH.

Root, A. I. and E. R. Root. 1923. The ABC and XYZ of bee culture. A. I. Root, Medina, OH.

Rusden, Moses. 1685. A full discovery of bees. Henry Million, London.

Simpson, J. A. and E. S. C. Weiner. 1989. Oxford English dictionary. 2nd ed. Oxford University Press, Oxford, England.

Shimanuki, H, K. Flottum, and A. Harman. 2007. The ABC & XYZ of Bee Culture. 41st ed. A. I. Root Co., Medina, OH.

Southerne, Edmond. 1593. A treatise concerning the right use and ordering of bees: newlie made and set forth, according to the author's owne experience: (which by any heretofore hath not been done). Thomas Owen for Thomas Woodcocke, London.

Tabraham, C. 1999. Tolquhon Castle. Historic Scotland, Edinburgh.

Taylor, H. 1860. The beekeeper's manual. 6th ed. Groombridge and Sons, London.

Taylor, H. 1880. The bee-keeper's manual revised by Alfred Watts. Henry J. Drane, London.

Thacher, J. A. 1829. A practical treatise on the management of bees. Marsh & Capen, Boston.

Thorley, John. 1744. Melisselogia or, the female monarchy. N. Thorley, London.

Tusser, T. 1586. Five hundred pointes of good husbandrie. Henrie Denham, London.

Walker, P and E. Crane. 1999. English beekeeping from local records up to the end of the Norman period. The Local Historian 29(3): 130–151.

Walker, P and E. Crane. 2001. English beekeeping from c 1200 to 1850: evidence from local records. The Local Historian 31(1): 3–30.

Walker, P. 1988. Bee boles and past beekeeping in Scotland. Review of Scottish Culture 4: 105–117.

Wheler, G. 1682. A journey into Greece. W. Cademan and others, London.

Wighton, J. 1842. The history and management of bees. Longman and Co., London.

Woodbury, T. W. 1862. Bees and bee-keeping; with an account of the introduction of the Ligurian species of honey-bee into Great Britain. Bath and West of England Agricultural Journal 10(1): 84–101.

Index